HEATH NOSTALGIA

A brief history of the Heath Company of Benton Harbor, Michigan - with fond memories of and by some of those responsible for the Heathkit name becoming world famous.

Published in Snohomish, Washington

This book is dedicated to the memory of Bjorn Heyning. Only one of his many contributions to Heath Company was his continuing effort to see that its history would not be forgotten.

Library of Congress Catalog Card Number: 93-93530
ISBN: 0-9637627-0-2

Credits

Thank you to all who contributed to this book by taking the time to write about their experiences, and to several who at least intended to.

Thank you also to Heath management for its permission to use company photos, logos, and documents.

Finally, thank you to all those people who made Heath Company such a pleasant place to work, who taught and guided me, who provided me with many enjoyable kitbuilding hours of my own, and whose friendships I continue to enjoy.

Terry Perdue

Contributors

Denton Bramwell
Chas Christy
Mike Elliott
Gene Fiebich
Bob Fernau
Earl Harris
Carl Helber
Bjorn Heyning
Larry Houghton
Randy Kaeding
James Lytle
Ed Mosher
Robert Paxson
Terry Perdue
Rich and Beth Renken
Neal Turner

Cover Design

Jannelle Loewen

Forward

Heath Company of Benton Harbor, Michigan had been producing Heathkits, build-it-yourself electronic products, for nearly a half-century. Most of Heath's customers found following clear step-by-step instructions to build a working electronic product from a box of individual parts to be an enjoyable, relaxing, and satisfying experience. A worldwide following of Heathkit fans looked forward to each new Heath catalog to see what new offerings they had to choose from for their next kitbuilding adventure.

Unfortunately, as this is written, the past tense used above seems appropriate, as Heath has abandoned all but a handful of selections from its once diverse line of kit products. When I joined Heath Company in 1973, the catalog contained some 350 different kits, nearly all of which were designed in-house. An equal number of assembled and accessory items was also offered. Heath's current catalog contains many assembled "buy-sell" items, but the few kits that remain are intended for the beginner. This seems a good time to put together a nostalgic look at Heath's past.

Since I'm not a historian, there may be pieces missing from my chronology. The parts I couldn't relate firsthand were obtained from others who could, and from magazine and newspaper clippings. Since most of this book was contributed by people who were with the company during the early years, their stories should help to fill in some of the voids I may have left.

My goal with this book was to paint a picture of the pride and dedication of the employees, suppliers and customers responsible for Heath's past success. I also hope to keep alive some of the memories, and to share them with others. Obviously, the stories contained herein represent only a small sampling of those that have been or might be told.

Finally, there are so many people who have contributed so much to Heath's success over the years; I truly regret that this book can't give all of them the credit they deserve.

Terry Perdue, K8TP

HEATH HISTORY

In the beginning...

In the early 1900's, young Edward Bayard Heath moved from Amsterdam, New York to Chicago, and started the Aerial Vehicle Company. Ed Heath, or "Eddie" as he was known, was an acquaintance of Charles Lindberg, and, like Lindy, had a keen interest in flying.

He also had a talent for light aircraft design. His company, later renamed Heath Airplane Company, became well known to flying enthusiasts across the country. Ed not only designed and manufactured light aircraft, but also provided a source of parts and accessories, gave flying lessons, and became an accomplished pilot. In the 20's and 30's he won many national competitions in his Parasol and Baby Bullet models, which were only two of many designs and variations. A fearless flier, he was the first to put a glider, his bi-wing "Super-Soarer", through a loop-the-loop; from an altitude of 1200 feet, he made four loops and landed within feet of his take-off point.

Heath sought to share his love of flying with others by making several of his planes available in a variety of kit configurations. Kits of parts for the Heath Parasol could be purchased for individual assemblies so that the builder could build, and pay, a little at a time. Assemblies could also be purchased pre-built, so that the customer didn't have to build the entire aircraft from scratch. One could purchase anything from full-sized plans for $99 to a fully-assembled plane, ready to fly, for under $1000, or anything in-between.

A series of articles describing the building of the Parasol was published in Popular Aviation magazine in 1930, written by Ed Heath himself. Modern Mechanics magazine ran a similar series by Stuart Rouse the same year. The Parasol model, so-named due to its high wing that acted as an umbrella for the pilot, had a reputation for being easy to fly, largely due to its short takeoff requirement and low landing speed.

Heath's first catalog was mailed out in 1925. It promoted his flying school, advertised planes, and listed parts, not only for his own planes, but for most models manufactured in those days. The Heath name was so well known that another aircraft parts supplier tried to take advantage by running ads under the name Heathe!

Ed Heath died in February, 1931, when an aluminum wing strut collapsed on a low-wing plane he was test-flying, and the plane dove 200 feet to crash on a farm near Morton Grove, Illinois, near the site of the present-day O'Hare. Soon after Ed's death, Walter Clinnin purchased the company from Ed's wife, and moved the business to vacated factory buildings in Niles, Michigan, near a large municipal airport.

Late that year, an improved model of the Parasol, the LNB-4, received an Approved Type Certificate, to comply with government regulations pertaining to unlicensed aircraft. Later designs, with larger engines and different wing configurations, were similarly certified.

In 1933 the company was reorganized and renamed the International Aircraft Corporation, although the planes were still sold under the established Heath name. Under Clinnin's direction, the company drastically cut advertising, discontinued the flying school, and was involved in some questionable dealings that eventally led to trouble with the IRS and the company being shut down in early 1934.

A new Heath Company is born

Howard Anthony was born in Dowagiac, Michigan on February 9, 1912. As a youngster, he became interested in electronics and flying, among many other things.

After graduating from Dowagiac High School in 1929, he spent two years at Hillsdale College in Hillsdale, Michigan, harvesting pickles in the summer to help pay his tuition. Following college, he operated a radio shop for several years while satisfying his other major interest by taking flying lessons in South Bend.

In 1935, Howard married Mr. Clinnin's secretary, and purchased, at auction, the entire company assets and the rights to the Heath name from the IRS for $300. Howard and his new wife Helen ran the business from Dowagiac, as Heath Aircraft Company, to take advantage of the good Heath name and avoid association with the bad publicity generated by International's failure. Together, Howard, with his technical talent, and Helen with her business savy, made a good team.

Within months of their purchase, the Anthonys moved their company into 4200 square feet of the former Baker-Vawter building on Park Street in Benton Harbor. They continued to sell the Parasol, and published the Heath Beacon, listing other flying-related items. With the slogan "Heath supplies the man who flies", the new company became known as a source for most any kind of light aircraft accessory. By 1936, Heath had also become the fifth-largest manufacturer of aircraft engines in the U.S. A new training program was begun, similar to that provided by Ed Heath's company, in which students paid for the necessary materials to build a plane, and did so under supervision of the company staff while learning to fly.

Howard recognized the need for a low-cost receiver for airplane use, and contracted with Meissner to design such a receiver to his specifications. This popular $39.50 unit was followed in 1944 by a companion HT-4 transmitter, designed by Southwest Airmotive in Dallas, and a number of accessories, one of which was a loop antenna for the receiver. Improved models were later designed in-house.

Several new products were introduced. One extremely popular item was the Heath Airphone, a simple stethescope-like affair that allowed an instructor to be heard by his student above the roar of the engine. The company also designed "lifetime" tail wheels for light aircraft, which sold for considerably less than those available from Goodyear. More items were added - non-yellowing acrylic windshields, pontoons, skiis, inspection covers, tie-downs, and more.

The company prospered during World War II, when it provided, to the armed forces, many items designed with non-strategic materials, such as pontoons made of plywood. During the war years Heath employed about 80 workers who labored long hours to meet the military contracts.

After the war was over, private flying was in a slump, and Heath ceased the sale of airplane accessories. Business declined, and the company shrank to a handful of employees. At the suggestion of some flying friends, the Zollar brothers, Howard began bidding on war surplus at government auctions. This resulted in several Benton Harbor warehouses being filled with literally millions of pounds of not only aircraft parts, but electronic components and assemblies as well. The Zollars took the aircraft parts and formed Aircraft Components, Inc., while Heath began to advertise the electronic items in a monthly flier, the forerunner to the Heath Catalog.

The items offered in the fliers were accompanied by reasonably-complete descriptions, and possible applications were suggested. In the case of assemblies such as receivers and transmitters, schematics were sometimes provided, as well as conversion information. Assortments of common parts such as resistors and capacitors were also sold. One flier apologized that Heath's resistor assortments had sold out, but stated that new assortments had been put together from resistors purchased elsewhere. However, it cautioned that since the new parts were not their own, "...we do not recommend your buying them."!

The first Heathkit

Before the war, Howard had suggested to Hickok Instruments the idea of selling test equipment in kit form, but they had no interest in the concept. With Howard's vast stock of components, he was now in a position to test the idea himself. One item he had many thousands of was 5BP1 cathode ray tubes. A friend suggested that it might be possible to develop an oscilloscope around this tube. Another company provided the actual design work, using five other common tubes, and the result was Heath's first electronic kit, the O-1 5" oscilloscope.

An ad was run in Radio News (later to become Electronics World), just to see if there might be a market for a kit scope at $39.50. The response to the test ad was overwhelming, and the Heath Company we've come to know "took off". The O-1 was soon followed by the V-1 vacuum tube voltmeter, then by the G-1 signal generator and other kit test equipment. (Model numbers were often omitted in ads in those days - there were still too few products to cause any confusion.)

Whenever possible, improvements were incorporated into existing products, and customers were either offered update kits or provided with modification information in one of their fliers. Customer's suggestions and feedback on products resulted in new and better designs, and Heath customers appreciated a company who listened and responded to their needs.

From the successful test instrument line, Heath branched into high-fidelity audio equipment in 1952 and, before long, followed with marine products and amateur radio gear.

As the product lines grew, so did the quality of the products and manuals. As the stock of surplus components ran out, new parts were purchased, tailor-made for each application. Since it was assumed that the builder would have had some experience with electronic assembly, early manuals were written by the design engineer, and somewhat crude, consisting of a few mimeographed pages of text (no step-by-step instructions!) and blueprints of the schematic and basic parts layout. Later manuals were expanded, and contained clearer illustrations of a step-by-step assembly process, and expanded circuit description, trouble-shooting and applications sections.

To aid in the selection of test equipment, Heath offered two free publications. **Planning Your Service Business** was written by John Frye, the author of the popular Mac's Service Shop and Carl and Jerry installments in monthly electronics magazines. **Establishing the Industrial Electronics Laboratory** was written by Louis Garner, Jr., another familiar name in the periodicals in those days.

To further enhance the usefulness of Heath instruments, the company made available, for only $2, a binder of technical application bulletins - about 200 pages describing various ways to get more benefit from Heath test equipment. Most of these were also written by Lou Garner.

Early-on, Howard realized the need to design kits in such a way as to allow customers to easily align them. The V-1 VTVM offers an example. To what standards would the average buyer be able to calibrate the finished meter? Howard solved the problem by measuring the voltage of the battery sent with each kit to supply power for the resistance ranges. That provided an inexpensive, one-time voltage calibration standard. For the A.C. ranges, he suggested that the builder set the meter to an average powerline voltage. Those customers who needed "laboratory" accuracy probably had access to more accurate calibration standards, but for many, the accuracy resulting from the simpler method was adequate.

In the case of the G-1 signal generator, Howard found a coil supplier who could wind the coils to such a tight tolerance that only one trimmer capacitor was needed; adjustment near the top end of the A.M. broadcast band resulted in excellent calibration accuracy throughout the range of the generator. (Many clever means have been devised through the years to allow customers to assemble, align, and maintain their products, a requirement constantly in mind throughout the development and manual-writing cycles.)

The relationships established between Howard and his suppliers was of mutual benefit on many occasions. There are stories about suppliers going out of their way when Heath was in a bind for parts to fill orders. There are also stories of how Howard insisted on paying a price that he considered fair, even if the vendor's quote was lower!

By 1954, the Heathkit catalog contained some 60 kits. It was about that time that Heath got into the amateur radio market with the AT-1 CW transmitter, AR-2 receiver, AM-1 impedance bridge, and AC-1 antenna coupler. (Actually, one could say that Heath entered the ham market almost a decade earlier, when they marketed war surplus communications equipment with conversion data. A simple one-tube, crystal-controlled CW transmitter concocted from surplus parts was also offered in those days.)

History repeats itself

On July 23rd, 1954, Howard Anthony joined four other passengers for a demonstration flight out of Ross Field in Benton Harbor in a twin-engine DeHavilland Dove he was considering buying, bound for Miami. Caught in a violent storm over Tennessee, the plane came apart and all on board perished.

Accounts of Howard Anthony by those who knew him often include words that exemplify the successful entrepreneur - instantly likeable, generous, enthusiastic, energetic, persuasive, resourceful, and honest. His fairness and concern for employees, vendors and customers is well documented. Helen Anthony certainly shared those qualities. The founding principles they set forth were surely responsible for the company's phenomenal growth in the decades that followed.

Following Howard's death, Helen Anthony carried on the company's operations while looking for a buyer that would assure the company's continued success. She was also determined that the company should remain in the Benton Harbor area, and that the current employees' jobs would always be assured. From among many bids, Daystrom, Inc. was selected for its similar business ethics, and Robert Erickson became Heath's new president in 1955.

Under the new corporate leadership, several changes occurred. A time-payment plan was established, allowing customers to afford some of the increasingly elaborate products offered. Daystrom began to establish Heath distribution centers outside the U.S., beginning with Canada, Germany and England. A network of Heath dealers was created in the U.S., beginning with one in Providence, RI., giving potential customers a chance to examine a completed product before purchasing the kit.

Heath's sales had increased many-fold in the years just prior to its sale to Daystrom. As a result, its operations had, for lack of space, spread out into five or six different buildings in Benton Harbor. By 1957 it was obvious that efficiency was suffering as a result.

Daystrom contracted for the construction of a new 150,000 square foot building on Hilltop Road in Benton Harbor's sister city, St. Joseph, and ground-breaking took place early that year. By February, 1958, all departments were back under one roof, (in the shape of an "H"), in a picturesque setting overlooking Lake Michigan.

The new plant was expanded several times in the following decade - by 60,000 square feet in 1962, and another 11,000 in 1967. In 1968 a computerized, multi-story "stacker" was added that allowed more efficient storage and retreival of the growing parts inventory. The plant expanded again in 1969, 1972, 1974, and 1980.

Despite the company's rapid growth, much of the "family" feeling inspired by the Anthonys was shared by Heath employees. The annual Heath picnics and Christmas parties were always memorable occasions. In the 60s and 70s, Christmas parties were held at the Shadowland Ballroom at Silver Beach, a few miles north of the Heath plant. This large domed wooden stucture, demolished in the late 70s, was once filled with the big band sounds of Lawrence Welk, Kay Kyser, Woody Herman, and other famous bands.

In late 1961, a special line of Heath/Daystrom kits was introduced that were essentially the same as their Heath counterparts, but had enhanced styling, packaging and manuals, and some pre-assembled boards to make them more appealing to retailers, since they weren't directly competitive with the mail-order products.

In 1962, Daystrom was acquired by Schlumberger Products Corporation. Schlumberger provided both the financial and technical resources to help Heath expand into new areas. Schlumberger also established a network of Heathkit Electronic Centers across the country, and established Veritechnology Electronics Corporation (VEC) to operate the centers.

Each new Heath catalog introduced new products in the existing product lines - instruments, marine, amateur radio, and hi-fi. Heath also added many new product lines through the years. Not all of them were related to electronics, and not all of them were successful, but Heath continually looked for new areas to enter. (A few product highlights are listed later in this book.)

As head of Engineering from 1954 to 1973, Eugene Fiebich was able to persuade management that Engineering could design, and Manual could clearly describe the construction of, increasingly sophisticated products. Under his guidance, the number of kit products increased from sixteen to over three hundred.

The birth of Zenith Data Systems

In the early 1970s, several companies were selling personal computers, still somewhat of a curiousity. Seeing an opportunity in this market, Heath went to work on computer systems of its own. In 1977 it introduced the H8 and H11 computers, along with several peripherals and courses. These systems were instant successes, and led to the Z/H89 in 1979, and PC-compatibles in 1984. A Heath User's Group, HUG, led by Bob Furtaw, was formed, and soon annual "HUGGIE" conventions drew Heath computer fans from all over the country. A number of small companies formed to furnish add-on cards for the popular Heath systems.

The computer business was different enough from the traditional Heath business, if that's definable, that it began to become a business unto itself, with separate marketing and other support groups.

In October, 1979, Zenith Electronics Corporation of Chicago purchased Heath for $64.5 million, to acquire its successful computer business. For several years, Zenith Data Systems was a separate entity, but shared facilities and resources with Heath. The "sharing" became more and more inequitable, however, and Heath products were often bumped as a result of "hot" computer products, resulting in schedule slippage. With its entry into PC-compatible systems, ZDS soon grew to match the size of the other development groups combined.

In 1988, a new building was constructed on the south side of Hilltop Road, facing the 30-year-old Heath plant. This was to become the new Heath Company complex, and the growing ZDS was to expand into the vacated areas of the existing Heath plant. Prior to its completion, however, ZDS claimed the new building for itself.

Heathkit's Decline

Somewhere in this time frame Heath management decided it could reduce overhead by moving kit manufacturing to the Far East. A small company in Hong Kong, Prokit, was formed, and it was expected that they would handle the sourcing and packing of parts for all Heath kit products, as well as some limited design work. They were also used to assemble, or farm out the assembly of, the Reflex home security lighting products. While these high volume items were produced much more economically there, problems caused by language and time differences, parts procurement, and priorities resulted in delays of as much as a year for lower volume products.

Late in 1989, financially-strapped Zenith Electronics sold Zenith Data Systems, Heath, and Veritechnology to Paris-based mainframe manufacturer Groupe Bull. The following year, rumors began to circulate that Groupe Bull was trying to sell the Heath business, which it never really wanted in the first place. Although the rumors were initially denied by company officials, it was announced in October, 1990 that Heath was indeed for sale.

Meanwhile, Heath management decided to drop most of its remaining product lines and focus on Education, Home Security and Home Automation. All kit products remaining in the Heath/Zenith stores across the country were returned to the factory and sold at discount in sale catalogs, at the Dayton Hamvention, and elsewhere. Many of the new product offerings were developed on the outside. The diversity that had helped to keep the business stable in the past was gone, and it appeared that few really unique new products would be offered in the future.

Early in 1991, Heath moved its operations from several locations in the twin cities area into a newly remodeled former K-Mart building in Benton Harbor. A manufacturing area was established in the new building for production of products that had proven to be inappropriate for overseas manufacture. With the elimination of the Heath engineering and marketing groups in late 1991, the need for in-house manufacturing is in question. Other groups were also severely cut, in a desperate attempt to break even by year's end. Unfortunately, it is a scenario that is all too common these days.

As this is written, in early 1992, Heath employs just over 150 people, and has not yet been sold. Direct mail and catalog sales have been discontinued. The Reflex (home automation/security) and education lines appear to be doing well, but the Heath Company that inspired such a loyal following for so many years is gone. Most of us who were part of Heath in its heyday are as saddened by the end of an era as by the loss of the jobs we loved. We know that many customers share our sadness. For us, the memories of the good years and the people we shared them with, both inside and outside the company, will live on. We would like nothing better than to see Heath return to the business and philosophy that made it so unique, once again giving people the pride and pleasure of building quality electronic products themselves. Unfortunately, people use their limited free time differently these days, and Heath has been forced to change its marketing philosophy accordingly.

The golden age of Heathkit is history, but some of the stories and pictures that follow should help to keep the memories of those days alive.

PICTURE POTPOURRI

Pontoons made the Heath Parasol at home on the water.

Edward Heath poses in his Baby Bullet.

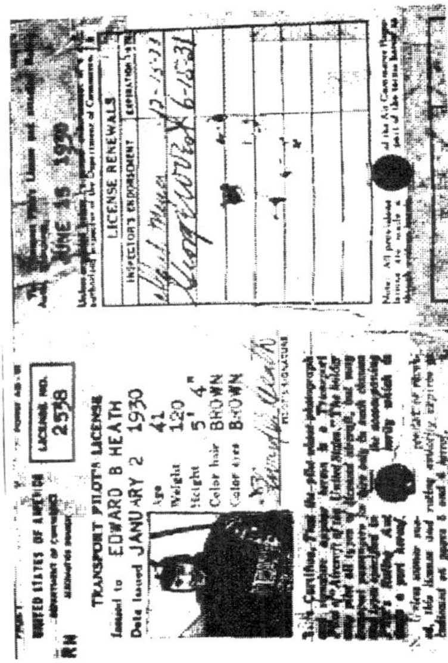

UNITED STATES OF AMERICA

TRANSPORT PILOT'S LICENSE

LICENSE NO. 2538

Issued to EDWARD B HEATH

Date Issued JANUARY 2 1930

Age 41

Weight 120

Height 5' 4"

Color hair BROWN

Color eyes BROWN

LICENSE RENEWALS

INSPECTOR'S ENDORSEMENT | EXPIRATION DATE

JUNE 15 1930

How to Build the Heath Parasol

Part IV

The Bolted Fuselage

Specially written for

POPULAR AVIATION

by

E. B. Heath

1888—EDWARD B. HEATH—1931

Edward B. Heath died on Feb. 1, 1931, as he would have wished to die—at the control stick of a new type of airplane which he was striving to perfect.

The airplane was an experimental ship of the low wing type, and was the first of its kind which he had built. A weakness in one of the wing struts developed while he was testing it, and the wing collapsed at a height of 1500 feet.

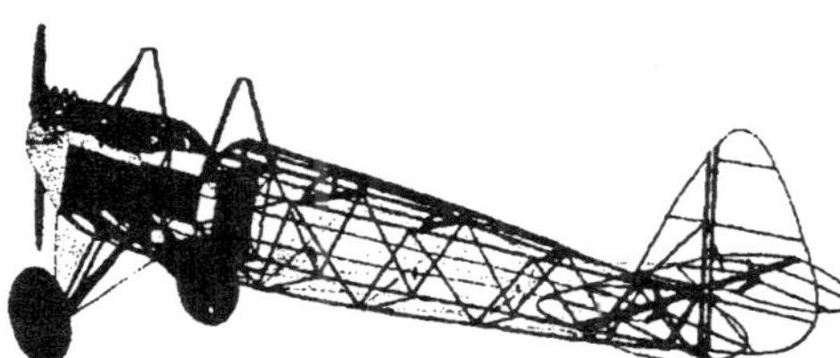

This view of the completed fuselage shows the simple bolted truss construction used. Note the minimum number of brace wires.

THE BOLTED FUSELAGE
by E. B. Heath

One installment in Ed Heath's series for Popular Aviation.

This three page advertisement also provided information on construction, pricing on subassemblies and plans, and general specifications. Kit price: $199, less engine.

Howard Anthony at his desk in 1950.

Howard Anthony
President, 1934-1954

Robert Erickson
President, 1955-1959

Daniel Knowland
Vice President, 1959-1959

Alan Greene
President, 1959-1963

Charlie Kirkland
President, 1963-1966

Dave Nurse
President, 1966-1980

Bill Johnson
Current President

Chas Gilmore
Current Vice President

One of the Heath Territorial plants- late 40s, early 50s.

The Hilltop Road plant that most readers will remember.

Heath's current home at 455 Riverview Drive in Benton Harbor was once a K-Mart. Extensive remodeling made it a comfortable home for the Reflex home security and Heath educational groups.

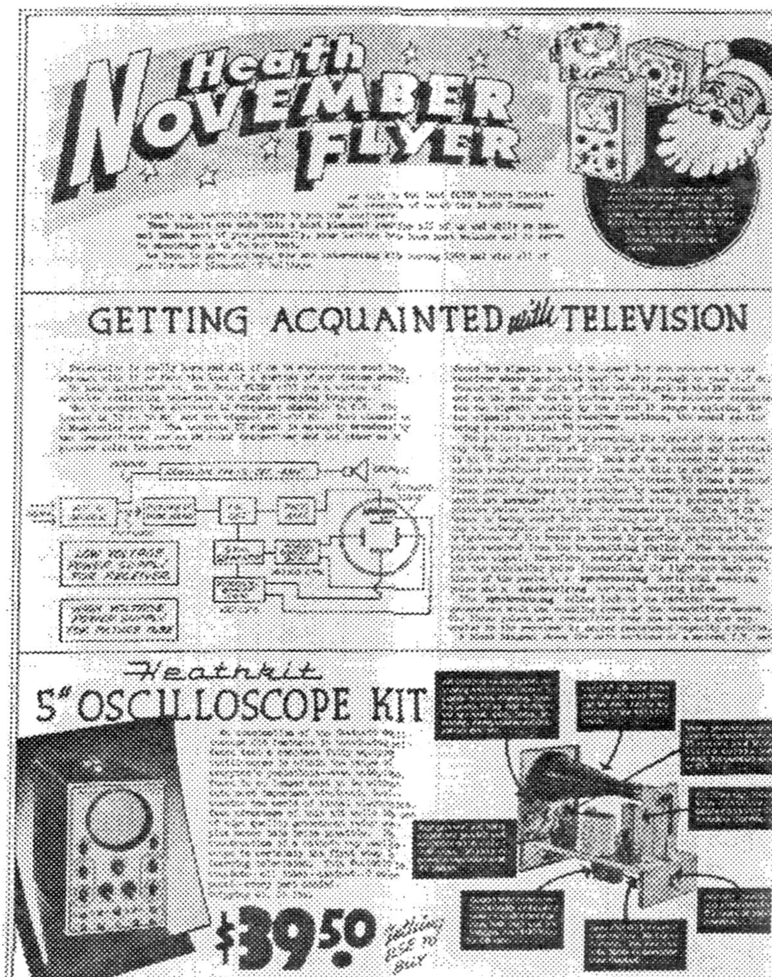

The cover of the four-page 1948 Christmas flyer. Hints and simple tutorials were common, as were cartoon-like characters.

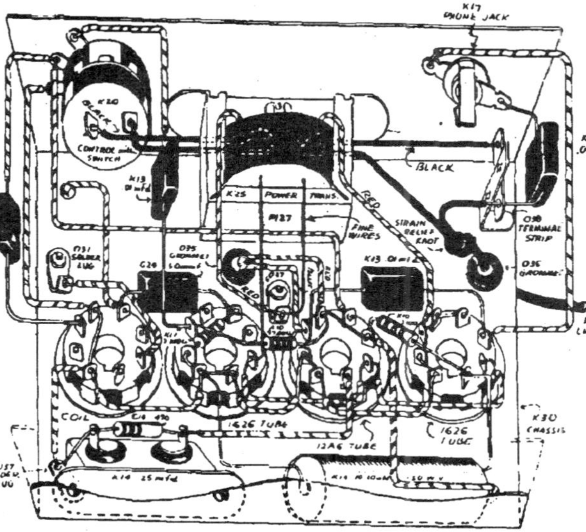

Heathkit

3 TUBE ALL-WAVE RADIO

MODEL K-2

The HEATH COMPANY
BENTON HARBOR, MICH.

6/20/49 K48

Even in the 40s, Heath's assembly manuals included clear instructions and pictorials to assure easy and proper construction. Today, many manufacturers of 'some assembly required' products still fall short in this regard.

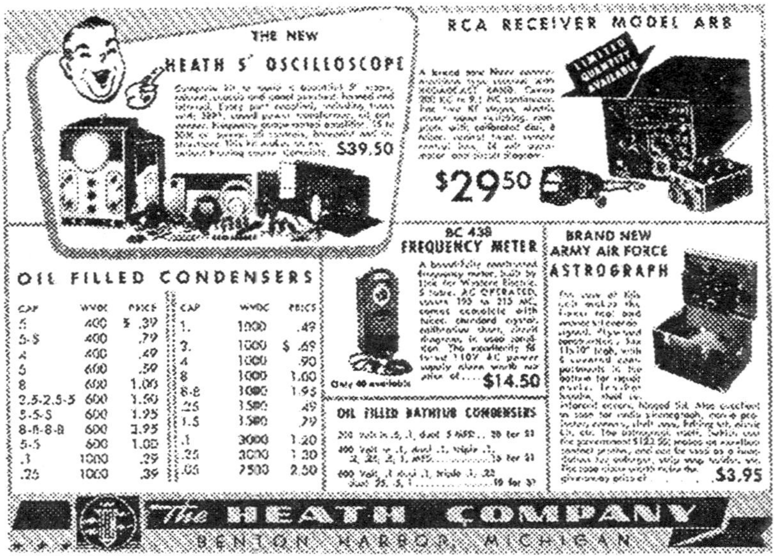

A typical magazine add from the 40s. Notice the combination of kits, surplus equipment and components.

This is believed to be Heath's first logo under Howard Anthony.

Early group shot of the Hams at Heath taken in front of plant at 305 Territorial in Benton Harbor.

In 1974, Heath hams presented American Radio Relay League station W1AW with a complete SB-104 station.

Design review of a new Heath amateur radio product. Left to right: Bjorn Heyning, Gene Fiebich, Bill Hannah, Carl Phlegar, Joe Shafer, Ray Freridge.

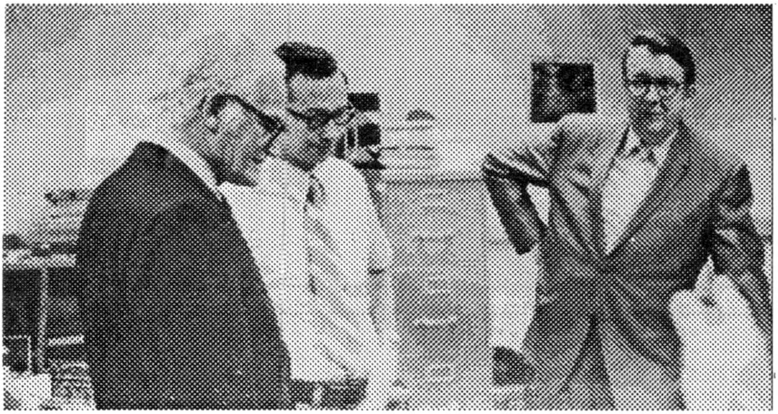

Senator Barry Goldwater, an avid Heathkit builder and amateur radio operator, toured the Heath Company in about 1972. Here he is viewing a prototype of one part of the SB-line of ham radio gear with Larry Houghton and Gene Fiebich. Goldwater built many Heathkits and was a booster of Heath wherever he went.

Design Engineer Reggie Sears confers with Sr. Design Engineer Roger Mace. The AT-1 transmitter at left was a popular starter rig, and the DX-100 was Heath's top-of-the-line CW/AM transmitter.

A well-outfitted lab at the Territorial plant.

Heath showroom at the new Hilltop plant in the late 50s. Amateur radio equipment occupies far corner in this view.

Hardware is carefully counted as bags are prepared.

Chassis are prepared for painting at the Hilltop Road plant. Installation of this Ransburg electrostatic system resulted in less wasted paint and more uniform coverage.

Computerized multi-story "stacker" at the Hilltop plant stores and retrieves pallets of parts automatically and makes efficient use of storage space.

Parts bags being hand packed in the Hilltop plant. These workers did their jobs with such care that it was very rare for a customer to be missing a single component, even in complex kits.

VTVM kits being packed at the Territorial plant.

A section of the metal shop in the Hilltop plant in 1957.

Subchassis being fabricated at the Territorial plant.

February 1, 1955

Dear Customer,

The final papers for the purchase of the Heath Company by Daystrom, Inc., were signed yesterday. I should not be honest if I did not admit some regret in relinquishing ownership of this company which my late husband and I guided from its start. My regret is lessened somewhat by the fact that I have great confidence in the new owners.

Daystrom is a management holding company which you will find listed on the New York Stock Exchange. Its subsidiaries are varied in nature with considerable weight in the electronics field. The oldest subsidiary is American Type Founders, which is known throughout the graphic arts industry for its presses and type supplies. You may be familiar with the Daystrom Furniture Division's dinette furniture. Daystrom Electric in Poughkeepsie, New York, does work for the armed forces and produces the Crestwood tape recorder for the consumer market. Daystrom Instrument does highly technical work for the armed forces and has recently introduced eleven instruments which have application in the atomic instrument field. Daystrom also controls Weston Electrical Instrument Corporation, which many of you may know for its fine precision measuring equipment. Daystrom-Pacific, recently acquired, is a lively young company in the guided missile field.

Daystrom has alert, progressive management. I chose them only after long consideration because I felt that, under their leadership, the Heath Company would continue to grow and improve. You, the customers, who have been our first concern will also be Daystrom's first concern. I hope the relationship will be as pleasant and profitable as ours has been.

Sincerely yours,

Helen C. Anthony

Helen C. Anthony

The general office area in the expansion on Territorial Road in Benton Harbor. Helen Anthony is at front left.

Early Purchasing Department at the Hilltop plant. The house visable through the window was on the south side of Hilltop Road, where Zenith Data Systems is today. The house was used as an employee salvage store for many years.

A corner of the Service Department in the 1950s.

Parts bags being packed for the DX-100 transmitter kit. Each bin is marked with the part number, description, and quantity required.

Heath contained a branch of the Benton Harbor Post Office in its Hilltop plant.

Design engineer Carl Heald works on Heath's impressive analog computer.

Industrial Designers like Stu Sizer made important contributions to product appearance and functionality.

The war surplus tubes that were used in thousands of VTVM kits were aged in these racks to assure stable performance.

The brilliant career of Howard E. Anthony, President of the Heath Company and originator of Heathkits, was tragically terminated July 23rd in an airplane crash which claimed six lives.

His record of accomplishment in the field of Electronics was particularly outstanding through his pioneering and complete leadership in the kit instrument field. His firm conviction in the feasibility of kit form electronics made it possible for untold thousands to benefit through the availability of equipment previously beyond their financial reach. Howard Anthony felt deep satisfaction in this personal accomplishment, and treasured the friendly encouragement of Heathkit customers above any monetary reward resulting from Heath Company operation.

Howard E. Anthony was a genuinely sincere civic leader, a highly regarded man in his communiity and was always ready to lend his support to any beneficial community enterprise. Loss to his immediate family, co-workers and Heathkit supporters is incalculable, but the Heath Company will continue to operate as a living memorial to his genius through perpetuation of his engineering ideals and business principles.

Report to customers of Howard Anthony's death

MEMORY MISCELLANY

This section contains stories submitted by various contributors. Many were past Heath employees; others were vendors or customers. With stories on various subjects from different perspectives, readers are bound to find something of interest here. While some of the anecdotes are of limited historical value, they were included for their humor. Note that some of the stories were written prior to this printing.

Remembrances I

When I joined Heath, the plant was in the lower section of the building at 305 Territorial, with some space also in the Baker-Vawter building across the alley on Park street. It was below street level. That had caused a problem the month before, when a heavy rainstorm had flooded Ox creek and water was standing over the curb on the street. Inside the plant the water had been over the desks in the office and phones had been hung from the heating pipes below the ceiling. Many of the files in the little file room had been stuck together and smelled terrible when unstuck.

My workbench was in that file room to start. Across the way was the office section with Howard and Helen's office. Cliff Edwards, Advertising, and Don Reagan, Sales mgr., in the next office. Then Gene Woods, Chief Engineer, and Chuck Christy, draftsman. Another cubbyhole for the Purchasing agent, Dick Cook, completed the private offices. A couple of girls had desks and there was a Switchboard with an operator. Behind the office section was the factory area, extending to the alley.

It contained the machine shop area, work benches with some foot-powered riveters for the inspection covers, the form blocks and the oven for the windshields, and the stockroom. Eldon Smith was the stockkeeper. Ralph Lhotka was Howard's right hand, and we had some fine people as crew, including a grandmother, who had her pet dog at her feet as she riveted inspection covers. There was a war on and help was hard to get. For coffee breaks we trotted across the alley into the Baker-Vawter building where the LaVanway's ran a refreshment stand.

Soon after I joined we had another flood, but with sandbags we kept the unhappy effects to a minimum. A faulty reverse-flow valve in the toilet near the alley let some evil smelling water into the factory area. We clearly had learned a lesson from the earlier flood.

For a while Howard had an assistant, but their thinking was not fully compatible and he soon left. We had a Test Pilot, Pappy Dillon, who had to try out the new designs, primarily the floats Gene Woods designed for Stinson L-5 and other small planes. They were test-flown on Austin Lake near Kalamazoo, but most of that was done before my arrival.

My task, besides checking out the Meissner MA-4A radio receivers before shipment, was to produce a small transmitter on the basis of a prototype developed by Howard's friends at Southwest Airmotive at Love field, Dallas as the Heath HT-4. It sported 4 tubes and transmitted at the then universal frequencies of 3105 and 6210 kc. Unlike the receiver, it required a 6 volt storage battery for power. We made several hundred of them, and I had an assistant, Wilma Felgner, who drew sets of parts from the stockroom, assembled, tested and packed them. Her record time was 4 hours for that process. We sold them to people with an A-2 priority for CAP/WTS (Civil Air Patrol/War Training Service) use.

The expanding volume of business let Howard enlarge the plant to include the upstairs, and soon new offices were built there. It must have been during that construction that I had my work space downstairs next to the spotwelder. A large sheet of cardboard kept the sparks from burning holes in my drawings. My old drafting table (which I still have at home!) served as my desk, my layout table and workbench as well. It was conveniently close to the drinking fountain and to the airhose, which I used to keep my pipes clean.

We made other interesting radio accessories at the time. A vital one was the Loop antenna, which illustrated Howard's genius. He knew that most loops had slip-ring contacts for connecting the rotating loop to the stationary base. They would get dirty and create noise in the signal. He hit on the idea of limiting the rotation with a small stack of dog-washers to somewhat over 360 degrees, which was enough to get the two nulls from the signal. Thus he could make the connection from loop to base with some very flexible tinsel wire, as used in shaver cords. It saved so much in the cost that he could sell a loop for less than a third of the current models. He had the base casting made locally and we cut the loops out of coiled brass tubing on a bandsaw. Then some drilling, brazing, plating, wiring and assembly made it a finished product, which worked faultlessly for many years.

Other radio products were a Trailing Antenna reel, made of scrap from the Plexiglas windshield material. And we made some metal pans for Storage batteries, which were heavily coated with acid-proof material. They kept acid spills from harming the plane.

The Meissner-made MA-4A receiver also reflected Howard's talents, for at his behest it used filament-type tubes, so it could operate from dry cells. The flat chassis was rubber-mounted to the main frame and case to prevent vibration from harming the fragile filament tubes. It was so compact that there was room in a small drawer under the radio for 3 D-cells and a 67 volt "B" battery.

When they ran down, they could be replaced from the cockpit - even in flight, if necessary. The A-model had the connections for the loop antenna and also a panel switch for fixed-tuned 278 kc Control Tower reception. It was quick and easy to install in the instrument panel, unlike most other aircraft radios, which needed a storage battery and a generator to charge it.

But it didn't have a Range Filter to suppress the 1020 Hz "A" and "N" signals from the Radio Range stations, when listening to the occasional voice signals on the same carrier. We made some experimental filters, but they never got into serious production.

Bjorn Heyning, 1986

Recollections of Howard Anthony

"Every person is unique, but some people are more unique than others", wise men say.

Howard Anthony was definitely unique. He was also enthusiastic, energetic, eccentric, excitable, volatile - altogether a doer with intelligence that would undoubtedly have placed him in the genius level.

During his high school days, Howard was one of a small group of friends who enjoyed each other and carried on together. They included Lawrence Brown, who was rather small in stature and served as the group's comedian, Fayette Howell and others, but particularly my next oldest brother Gordon Paxson. He and Howard shared numerous adventures through the years.

Many of my recollections of Howard are First Hand, which I will designate with (FH); others I heard told in good faith by Gordon or our older brother Howard Paxson, who also spent much time with Anthony. These yarns I'll mark (HS) for Hearsay.

You can question them if you wish, but I believe them. Unfortunately, those who could corroborate them seem to have left this "mortal coil".

1. THE FIRST ANTHONY RADIO FACTORY (HS)

Radios, during the late 20's and early 30's were undergoing rapid technical advances. Instead of power from unsightly big batteries they were improved to accept 110-volt power from any convenient wall outlet. While in high school, Howard determined to take advantage of this burgeoning market. He designed a chassis and circuit, enclosed it in a round-top wood cabinet about a foot high, got together enough parts for a small order, and with a friend assembled a dozen or so sets on his kitchen table at home. We acquired one of the radios. The set sold for $9.95.

After they had sold all their production, they discovered they had lost money on every set. A businessman's first lesson.

2. TWO 1918 DODGE TOURING CARS (FH)

After graduating from high school in 1930 my brother Gordon bought a decrepit black 1918 Dodge Touring car for $25. The Dodge was built tough as a tractor. It beat walking by a mile. On or about the same time Anthony drove up with an identical old black Dodge Touring car. The young guys were delighted to park their cars side by side in our front yard, a few feet from our porch, and admire their wonderful possessions, each bragging about the superior performance of his own car.

One day Anthony drove up, parked his car beside Gordon's, (which they had dubbed "Ahab") and showed Gordon a huge dent someone had jammed into the driver's door of his car. They both laughed at this insult - until Howard walked around to Gordon's car and suddenly hauled off and KICKED a huge dent into the door of Gordon's car. "Now they're alike again", he yelled.

Whereupon both broke up into a laughing, howling fit.

3. A FIRST AUTO RADIO (FH)

Sometime later, Howard drove into the yard in the old Dodge. Gone was the back seat and rear of the body. He had cut it off and built a wooden pickup body to replace it. Over the driver's seat was a new wooden top, and behind it a large black trunk.

"Get in", he urged us. Gordon and I got in the front seat beside him. Reaching over to a new group of instruments on the dashboard he flicked a switch. Out blared MUSIC from a Chicago radio station! Amazing! A radio in a CAR? "Will it play while you're driving?" "Of course", he grinned.

He had installed a tuner and speaker up front and the chassis and big batteries in the trunk behind. It was incredible - the first radio we'd ever heard playing in an automobile. And Anthony had built it.

4. THE PICKLE PATCH (FH)

After high school, Howard attended Hillsdale College for a year. While casting about to earn money during the summer, he must have heard that farmers the previous year had made good money raising cucumbers. So he and, I believe, a friend rented several acres from an uncle's farm near town and planted cucumbers. When the plants began to bear, he hired another kid and me to help harvest them.

So he'd come by with the old Dodge, now called the "Pickle Truck" and take us out to the "pickle patch". Stooping, or on hand and knees, we'd fill baskets with cucumbers all day. The vines are quite prickly and after a rain it was muddy work. It wasn't a fun job. But there was a lot of constant joking to shorten the day.

After loading the truck, we'd take the cukes down to the railroad depot and send them off.

Howard apparently earned enough to return to college in the fall, and I had some cash for school clothes. A good deal! Howard would try anything legitimate to make a buck.

5. REBIRTH OF THE HEATH COMPANY (FH)

Others have told the story of how Anthony acquired the assets of the defunct Heath Airplane Company.

About seven years had passed since the pickle patch, and Howard had established the Heath Company in a Benton Harbor building on Park street.

Howard and Helen were now having a terrific struggle making ends meet. With classified ads in various aviation and home mechanics magazines, they offered literature describing the Heath Parasol - "Send one dime (10 cents) for literature." Each morning they went to the post office, shook out the dimes and had breakfast on the day's receipts. They lived in one rented room. Helen was quite a lady.

They also sold off the stock of Heath parts: Some remaining Henderson engines, replacement valves, kits and other parts to keep the company doors open. But as the old stock dwindled, something had to be done.

With his typical ingenuity and inventiveness, Howard got a mold together and had a local rubber company start turning out replacement light plane tailwheel tires. At the time, Goodyear was marketing tires at about $5.95.

By this time I was working as a copywriter for my brother Howard in Paxson Advertising agency in Benton Harbor. I was assigned the job of writing a one inch x one column ad for the tire - Howard Anthony's first ad with a tiny picture in it - and the product was priced at an astonishing $2.00! It was a big success.

Soon Howard added a complete tailwheel assembly - yoke, shaft, wheel and tire - designed to attach to the usual tailskid on Piper Cubs, Taylorcrafts, Aeroncas and other light planes. Another success!

From then on, one new product followed another. Many new pilots were gaining their wings in Piper Cubs, but the instructors had problems shouting their orders to students over the roar of the engines. So - dodging the then costly electronic solution of a radio type intercom, Howard produced the Heath Airphone. A rubber speaking cup for the instructor connected by a long rubber tube to a pair of rubber earphones for the student did the job quite adequately. The product served a real need - and sold by the thousands.

Howard's ads grew larger and his mail order business prospered. New products followed: Replacement windshields, tie-down kits, light plane skis, and finally a neat, small radio receiver, powered by a battery that fitted into a handy drawer in the cabinet. Again, this product far undersold others on the market.

By 1941 Paxson Advertising was placing Heath ads in magazines, including Aero Digest, Aviation, Popular Aviation and Air Facts. Most profitable of all was Trade-a-Plane News, a large yellow sheet that brought in continuous orders with ads measuring 7x17 inches and showing up to 18 different products.

Throughout the growth of the Heath Company, before and after this time, great credit is due to Howard's wife Helen for the firm's success. Her steady hand on the finances and the office staff kept the bills paid and the discounts taken. She also exerted a firm influence on Howard when she judged some of his wilder ideas were too far out. She was an invaluable part of the team.

6. SPEED DEMON (FH)

Before I got my own car, I rode to Dowagiac several times with Howard and Helen to visit our respective parents. Howard had a newish Ford V-8, and kept my hair standing on end as he wove the winding roads at 70 to 80 miles an hour. We weren't flying, but we barely touched the road. Riding with him was a thrill to remember!

7. HOWARD'S PARASOL (HS)

During his earlier years as Heath Company's owner, Howard had acquired a complete Parasol. How he learned to fly it I've never heard. Sometimes he flew the little ship to Dowagiac, landing in a pasture used by Jenny-flying barnstormers and others. Once I went out and met him there, and he took my picture climbing out of the tiny cockpit, although I never touched the throttle. Later I heard he had cracked it up and got out with a broken jaw. That's damage control!

We were amused to hear that a man who had bought a Parasol - in Colorado, I believe - failed to make his payments. So Howard drove out in his V-8 Ford to repossess the plane and started to trail it back behind his car. His well-known driving speed, however, was too much for the plane's little gear. His speed burned the tires right off the wheels. Somehow he retrieved the ship later.

8. WAR TIME (FH)

By the time the Japanese bombed Pearl Harbor in 1941, the Heath Company was really swinging. I had fun working with Howard. He helped keep me busy changing and rewriting ads to include new products every few months; often they were last minute changes, typical of his frenetic activities.

As an avid airplane model builder and aviation bug at age 14, I enjoyed having contact with the aviation business. This ended when I joined the army in January 1942.

In the Service as a medic, I worked three years in charge of medical dispensaries, then volunteered for the Air-Sea Rescue Service. As a medic-observer aboard a PBY Catalina amphibian, I finally realized a dream of flying in the Air Force. A lowly aircrewman, but flying. At last a "flyboy!" I could look down at the rest of the mundane world flying over the Pacific ocean from Okinawa.

9. BERTHA'S HOUSE (HS)

It may have been during the war that Anthony moved Heath into a newish cream-colored brick building on Territorial street. Next door was an infamous house run by Madam Bertha Russell. Many prominent residents were said to be frequent clients of her girls, but during the war a visit to Bertha's was highly popular with soldiers from Fort Custer and other military posts.

As a neighbor, Howard may have objected to her operation as

being bad for the neighborhood, at which Bertha is said to have replied testily: "You think you're helping the war effort? Well, my business is just as important to the war effort as yours is any day!"

Later, the State Police were called in; it was time to arrest Bertha and put her out of business. Her house was vacated. So Howard offered his staff the opportunity of touring the place before it was torn down. En masse, a group of men from Heath toured the house. As they came out like a clutch of tourists, Howard stood in his office window and filmed them all with a movie camera. At the next Christmas party for employees and their WIVES, Howard put up a screen and showed his movie of their husbands trouping out of Bertha's house. They say it was a hit film. Guess who got hit?

10. WINNING ON WAR SURPLUS (HS)

During the war, Howard and Helen continued building the Heath Company with government and private orders for original and replacement parts. When the war ended, orders for parts dropped precipitously. Howard scrambled. Hearing of a war surplus equipment sale, he bid and won a carload of unknown crated material. It was a chest of buried treasure.

Among other things were two powerful radio transmitting stations, a large quantity of new typewriters, a bunch of Astrographs (navigation instruments for pilots), and foretelling the future, a large quantity of radio and radar tubes and parts. Putting his ingenuity to work, Howard assembled enough parts to introduce the first Heath electronic kits.

The Heathkit business was born. May it continue to prosper!

11. THE WAYS OF BANKERS (HS)

My brother Howard Paxson tells of an early time when Anthony was in serious need of a bank loan. When asked for collateral, he could offer nothing the bank would accept; they turned him down. So my brother, who knew the bankers well, returned to the bank to vouch for Anthony as a good credit risk.

Years later, looking for a possible profitable investment, Howard risked some funds in a wildcat oil drilling firm. It paid off with several gushers! Now "in the money", Anthony was invited - and became - a Director of the same bank!

12. BACK TO HOME BASE (FH)

Returning to Paxson Advertising after the war, I found that another man had taken over Heath advertising. But I continued to see Anthony from time to time. Six post-war years had passed; for the last five Gordon had worked as Assistant to the Dean of the art department and taught at Syracuse University.

Anthony had felt a need for some expert industrial design work on the many new kits he was selling, and offered Gordon the Heath account as a start, if he would return to Benton Harbor and open an independent industrial design studio. Gordon accepted the offer, returned to this area to serve Heath and soon acquired the V-M Corporation and others as clients for his design services.

13. POST-WAR ADVENTURES (FH and HS)

About 1951, Anthony and my brother Gordon together ordered a carload of mahogany lumber to use in building a new home for each family. Gordon designed his own home and hired a contractor to build it. Howard admired the designs of famous architect Frank Lloyd Wright and hired him to create and supervise his new home on Miami Drive.

One day I met Howard in the post office. He was just opening a set of house plans received from Wright. He was excited and said: "These are beautiful! You should get Mr. Wright to design a house for you. His fee is just 15%, like any architect." I said, "Yes - but 15% of WHAT?"

During construction, his house kept increasing in cost from day to day. Then Wright took a trip to the Southwest. Soon a large package arrived C.O.D. to Anthony. Wright had sent a huge Mexican vase he decided should be in the house. More C.O.D. packages came - and more. Wright had reserved the privilege of selecting everything that went into his designs - expense was no object. The Anthony's were in the grip of a genius. He also designed all the furniture for "Scherzo", the name he chose for the house to match Anthony's quick, energetic personality.

At the rear of the home was a laboratory room for Howard's experiments. Gordon was visiting one night when Howard asked, "Have you ever heard a ship's radio talking at sea?" Gordon said he never had. As they visited, Howard threw parts together and shortly plugged in a radio chassis, tuned in, and there they were - listening to ships at sea! Nothing to it for Howard.

14. SAILING IN A GULF GALE (HS)

Sometime later Howard had bought the BETSY, a 32-foot auxiliary ketch. With some friends he sailed it down the Mississippi river to New Orleans, but his crew returned to

Michigan.

A phone call brought Gordon, who had some sailing experiences, down to join Howard. Sailing through the Intracoastal Waterway was peaceful and pleasant. But soon after entering the open Gulf of Mexico, an unpredicted high gale storm hit the BETSY. It was touch-and-go for the two sailors for two horrible days and nightmarish nights - a story better told in articles Gordon wrote for yachting magazines.

When they arrived shakily in St. Petersburg, commercial fishermen were aghast. None of them had dared venture into the Gulf for days!

15. VACATION TO JAMAICA (HS)

During these years, Howard, Gordon and Judge Tony Westin decided to fly commercial airlines to Jamaica for a vacation look-see at the Caribbean. It was a happy trip and while there, they all hopped over to Haiti. The terrible poverty there was brightened some by the large number of excellent artists, striving painfully to make a living painting. All the men bought some lovely paintings, partly to encourage the desperately poor artists. Howard was especially generous in his purchases.

16. THE LAST FLIGHT (HS)

On a flight to New York in Howard's Aero Commander, a Heath employee became sick, when they hit rough weather. As the Commander was not pressurized for flight over 10,000 feet, they couldn't rise above the turbulence.

Howard decided to consider buying a larger, safer, better-equipped aircraft and try a test flight to Florida, after which he and Gordon would fly commercial to vacation again in Jamaica.

On the morning of July 23, 1954, Gordon Paxson's 42nd birthday, pilot Gordon Wyrick of the Gordon Air Service, Pontiac, MI, landed at Ross Field in a twin-engine DeHavilland Dove to demonstrate the aircraft to Anthony.

With Howard's corporate pilot, Lawrence Durham, at the controls, they took off with passengers Howard Anthony, Mr. and Mrs Laurel Brown and Gordon Paxson. Later a radio check by pilot Durham was made one hour and 12 minutes en route. The last entry in a diary kept by Mrs. Brown recorded that they were flying at 13,000 feet.

Eyewitnesses on the ground next saw the aircraft coming out of a stormy, rainy overcast at about 3800 feet, out of control in a steep spiral or flat spin. Witnesses then described a thunderous noise that came from the plane, and saw objects falling from it, prior to contact with the ground.

The aircraft struck the ground vertically in a mountainous section of Bledso county, Tennessee, exploded and scattered wreckage over a wide area. The right wing, both engines and other parts were found some distance from the remaining wreckage. All persons aboard were killed instantly.

Before and after the accident, there was heavy thunderstorm activity in the area. I have heard that this mountainous district in Tennessee is known to have severe air turbulence during storms, which has caused numerous accidents to flyers.

It is my opinion, not substantiated by the Aviation Safety Agent's report, that the DeHavilland Dove encountered a devastating wind shear or downdraft over the mountains, which bottomed out so suddenly that it tore one wing (with engine) off the plane, rendering it totally uncontrollable.

The accident was a tragic end to a vacation for six valuable citizens in the prime of life.

With Helen's gallant management after the tragedy, the Heath Company endured.

Robert Paxson, 1989

Fond Memories of "The Good Old Days"

As a small, but growing, company, many situations occurred which could not be part of a corporate structure. For example:

Howard Anthony personally opened all Monday morning mail, the busiest day of the week. He was amazed at the steady weekly increase of orders, jumping from 500 to 1000 - to 1500! He was so enthusiastic that one Monday morning he loudly exclaimed: "If we ever hit 2500 orders, I'll treat the whole plant to a steak dinner!" About three weeks later the target point was quickly passed! Without further ado Howard's secretary, Lois Rudduck, immediately contacted Tosi's restaurant and made the arrangements. Tosi's was a small restaurant in those days and couldn't handle the entire group at once. The steak dinners were held on two consecutive Wednesday nights. Certain conniving employees managed to go to both of them.

Another example (one beautiful spring morning in 1952): We were firmly entrenched in our new building location (the former Bertha Russell whore house). Just before noon lunch break, Howard spoke over the P.A. system. He commented on the beautiful spring, fruit trees in blossom, grass turning green, birds singing, etc. He said it was too nice a day to work indoors. Everyone should straighten up their work area, leave for lunch, and not come back until Monday morning! Being Friday, this gave us all a nice long weekend with pay! We had 150-175 employees then. Just another pleasant memory, even if it was paternalism (which is not all bad).

The company family picnics and Christmas parties were legend, never to be forgotten.

Gene Fiebich, 1991

Random Ramblings about the Old Heath Company

In 1948 I was working for a Chicago electronics manufacturer, coordinating Military Contracts. At that time most of the post-war military contracts permitted the use of war surplus, provided certain conditions were met. And I had been told by some of my Chicago connections that "this guy Anthony" had quite a stock of surplus in Benton Harbor.

To make a long story short, I wound up with Ralph Lhotka going through some of the Heath warehouses to see exactly what was available and whether or not it would have any interest to my company. After a trip through several dingy buildings, Howard Anthony appeared on the scene, introduced himself and said, "I want to take you for lunch today and we are going to eat at a place that has the best chocolate pie in the world." First we took a short tour of the area, and as Howard was accustomed to do, it turned into a sort of Cook's Tour of the Benton Harbor - St. Joe area. I was really impressed with his enthusiasm about where he had located and the many advantages of the Twin Cities. Quite frankly, after having lived in Chicago for seven years, the advantages of a small town were very impressive to me.

I had to agree about the chocolate pie! It turned out to be Ma Brown's place out on Pipestone and we went back there many times in the following years to enjoy the food. During the luncheon, though, I mentioned to Howard that I had seen a stock of generators in his surplus that I knew were going to be needed by an army contract. He said, "Why don't you keep your eye on it? Let me know who gets it and maybe we can get together." That ended the discussion. I went back to Chicago, back to work and forgot about it.

After several months, the contract we talked about was awarded to a New York manufacturer and, remembering my conversation, I called Howard, gave him the name of the company etc., and he thanked me very much and that ended that.

The next contact with Heath was a complete surprise, because, without any forewarning, I received a check from Heath Company. This struck me as kind of a strange operation. I called Howard and he said, "I told you that if we could get together on this thing, we would make a deal. And we are going to get together. I would like you to get over here this weekend, if you can. These people are coming in and I may need your help." After a discussion with my present boss about a conflict of interest, I decided to risk it, and made the trip to Benton Harbor and sat down with Howard and the representatives from the successful bidder.

The haggling went on most of the day and I was rather bored with the whole thing. I had a lot of work to do in Chicago, and it seemed like a complete waste of time to me. Incidentally the amount of money involved was quite substantial. I had had no particular financial concern, thinking that what I had seen was my end of the deal. Typical of Howard's approach to the problem, I remember him saying later in the afternoon - as I remember this is exactly what he said: "Gentlemen, in 15 minutes I have a date to meet my wife downtown for dinner - in 15 minutes the price of this item is going to go up by 20%." He turned around and walked away from the meeting. The representatives from New York got on the telephone and made some very frenzied phone calls and Howard came back to the meeting - I was not in on that part of it - I don't know what happened, for I went back to Chicago and forgot the incident.

Another surprise! Shortly after that I received another check, with no explanation from Howard, which represented a commission on the purchase of the stuff by the New York concern. The commission check was considerably larger than it should have been, based on the traditional scale being paid on commission sales. This was rather impressive to me. I discussed the matter with my boss in Chicago and decided I might have a better financial future in the surplus business than in what I was doing at that time.

So I resigned, and again to make a long story short, started about a one year operation working with a Chicago surplus concern, selling to and buying from Howard, and becoming more and more impressed with his operation.

In May of 1949 we moved to Benton Harbor. It seemed that we had just barely gotten in ahead of the wire, because Heath Company's business began to grow at an unbelievable rate. I went back and looked through some of my old statistics and uncovered some pretty startling figures. For a four year period, ending about 1955, the company grew at the rate of 80 % per year. Expansion at that rate created a lot of problems, not the least of which was that of space. By 1954 we were running out of places to put people, stock and, frankly, money!

I had started out as sort of a coordinator of government contracts for Heath Company, but I wound up one morning being asked by

Howard if I would like to take a chance and write a couple of manuals for him. The work that I was doing didn't take up all of my time. I said, "Sure, I'll be glad to." I wound up as a design engineer, which was certainly outside my educational background. But I took on the chore and really enjoyed it.

Finally, after three years of miscellaneous manual writing, design engineering, etc., Howard asked me if I would take on the design of the new color T.V. oscilloscope. This was early in 1954. Technically, this was a bit beyond me, but I worked with Bjorn and with the other engineers and with Howard and did pretty much what I was told.

I remember very vividly sitting in my lab one Friday morning in July of 1954, when Howard came in with his usual windy, brusk appearance and dropped on my bench a prototype of a 5" oscilloscope. He said, "Turner, there it is, it's all done, it's designed, it works. All you need to do is to put it on printed circuit boards." And he turned around and left. He said, "I am heading for Florida, I'm going to look at a new airplane and I'll see you when I get back." That was the last time I ever saw Howard.

After his tragic death, Mrs. Anthony ran the corporation until January 1955, when we were told by her that she had agreed to sell the company to Daystrom, Inc. A condition of the sale was that the company had to stay in Benton Harbor and the present key personnel had to stay with it. Without revealing any confidential data, I think it is interesting to note that the gross sales of the company, in the following three months, were more than enough to cover the cost of the company to Daystrom.

A second personality, who had great influence on me was Bob Erickson, who negotiated the purchase of Heath by Daystrom. Bob, who soon became known to us as "Rapid Robert", because of the way he spoke and the way he moved, was a very dynamic guy. He introduced a lot of new business techniques, which sort of took the biting edge off of the rapid expansion thing. One of Bob's favorite comments was: "Remember, the average living room in the United States takes a 9 x 12 rug." Another comment, which bore directly on our problems: "There are no inventory controls as effective as four tight walls."

Bob and his wife Sally seemed to fit in very well with the Heath Company group, despite the fact that they were obviously from an Eastern establishment background, which, you would have thought, would have created some real problems in Benton

Harbor. But their personalities were such that, if that problem did exist, we certainly weren't aware of it.

The company continued its growth at a rate that would have been unbelievable, had you not actually lived through it. I remember that before Howard was killed, he stopped me one night, as I was leaving the plant and said: "Turner, if things continue the way they are, we are going to do five million dollars this year." Surely, after Daystrom bought Heath Company, we exceeded that number by an order of magnitude.

One of the most rapidly growing segments of the business was the hi-fidelity product line. At that time a conventional way to market Hi-Fi was by introducing products at so-called audio shows. The hi-fidelity manufacturers had gotten together into a trade association, to which they refused admission to Heath because we were "not manufacturers." Therefore we were not eligible to participate in these audio shows. But Bob Erickson was not going to be stopped by that technicality.

In February 1958 the hi-fidelity show was going to be held at the Ambassador Hotel in Los Angeles and we were not permitted to exhibit. So Bob set up a deal at the Alexandria Hotel, directly across the street from the Ambassador, and we sent out our audio product line and personnel to the West Coast. To say that the Heath show was a success would be an understatement! The crowds at the Alexandria became so large that the fire department refused admission to a long line of persons waiting to get into the show. Finally we had to actually go back down the line of people and tell them that they were not going to be permitted to come in that night. This continued for about three days, as I remember it. One of the crowning touches, that was typical of Bob Erickson's approach: we put a little sign at the entrance to our show, saying: "Please do not miss the hi-fidelity show across the street at the Ambassador."

This kind of marketing and this aggressive approach to problems were typical of Bob. They made an impression not only on Heath Company but also on potential buyers in the hi-fidelity and amateur radio fields. By 1959 Heath Company had at least 12 kits in the line that sold over 10,000 units each. In December 1958 we reached our peak. We handled 54,776 orders in the month of December. In the 24 months ending in the fall of 1959 we averaged over 28,000 orders per month.

In direct violation of Bob's own requirement about the four tight walls, it became obvious that we had to get to a bigger manufacturing facility. After a long, protracted zoning fight, and some discussions with the business communities in Benton Harbor and St. Joe, (and with the post office people) it was decided to move Heath Company's operations to St. Joseph. The new plant Open House was held in August of 1958, and at almost exactly the same time, Bob Erickson tendered his resignation from Daystrom.

I was fortunate (or unfortunate) enough to be with Bob at a meeting at Murray Hill, at which Bob made the third of his most impressive comments which I remember so well. There had been a long, heated discussion about how Heath Company had been handling its inventory of surplus cathode ray tubes, which still was quite large. And the Chairman of the Board of the Corporation asked Bob: "How could you possibly have made this calculation, which we find now to be rather misleading?" Bob looked him straight in the eye and said very simply: "I was wrong." That ended the argument and the discussion.

I don't know if this had anything to do with Bob's decision to resign or not, but he left us shortly after that to go to work for Beckman in California.

And that led me to the fourth of his comments, which I remember so well. He called me from the airport in Chicago one day and said something about buying some Beckman stock. He finally said, "Turner, let me tell you something I never want you to forget. When you change jobs, never move any farther East than you are." Bob now has resigned from Beckman and he and Sally, as I understand it, have spent many, many months traveling and playing golf, which are their principal activities.

A few interesting little anecdotes are probably worthy of mention. It was the custom at Christmas time for the employees to leave the neckties off on the last day of work before the Christmas holiday. If you forgot, you were apt to have your necktie sliced off neatly by some of the more active engineers and others. On one such day, Erickson and I were walking across Territorial to the plant, where the lunch was going to be held. I had forgotten about the fact that this was Christmas Eve. As we walked into the building, Carl Heald sneaked up behind me and with a large pair of tin snips removed the bottom 70% of my necktie. This was Erickson's first experience with this type of thing. He was quite shocked, and I was a little bit put out, because the tie was a gift from my son.

The first Christmas in the new plant, I decided that it was time to even the odds a bit. I made arrangements with Gene Fiebich to have Carl come to his office while I was there. Watching carefully, I waited until Carl left his lab, went in and got his necktie from his coatrack (because he had remembered the tradition). I put it on, went in and sat down by Gene at the conference table in his room. As Carl came in, he took one look at me, turned around, went out and said, "I'll be back in a minute." He came back behind me, reached down and nipped off the tie. And as he did so, he said, "That's funny, I've got a tie just like that one at home." Well, I guess that settled the score.

In conclusion, my 25 years at Heath were certainly most interesting and entertaining. They exposed me to two men, Howard Anthony and Bob Erickson, who not only were good friends and good business men, but the type of entrepreneurs who we need more of today. I sometimes wonder what Howard would think if he came back and saw the present Heath operation. Somehow, I think he would be disappointed.

Neal Turner, 1986

My Heath Story

I went to work for Heath Co. in August 1942 after graduating from High School. I had 4 years of Mechanical Drafting, so I became the first draftsman at Heath in Benton Harbor. I made all the drawings for the many accessory items for aircraft and also the installation drawings. As we received orders from the Army Air Corps, our own drawings had to be made from the Army blueprints. We worked on the light aircraft the Army would fly in for Heath to use for test purposes of floats, skis, windshields, tailwheels, etc.

Heath was provided a Stinson L-1 Observation type aircraft, to consider building the top canopy of Plexiglas. The canopy was quite large, about 5 feet wide x 10 feet long. In order to get measurements to make drawings for making a sample unit, some men from the windshield department were sent to the airport to take the canopy off and take it to the shop. It was cold and windy (which the airport was noted for) and the wind caught the canopy (which was shaped like a giant turtle shell). It bounced twice and broke into a thousand pieces! It took 2 or 3 days to tape it back together so we could measure it. After all that, we didn't get the job.

Another time we were provided with a factory-fresh 100 hp Piper Cub, to use for float testing. It was tied up on the beach at the Yacht Club basin by Lake Michigan. Test pilot Al (Pappy) Dillon and engineer Gene Woods decided to make some flight tests on a beautiful afternoon. They taxied into the basin and as they were making pre-flight checks, the nose went down and it sank in 6-8 feet of water. A boat was sent to pick up Al and Gene, who were on the top of the wing.

Howard had everyone in the shop and office (about 20 people) go to the lake, where they tied a large rope to the aircraft and pulled it up on the beach. It was discovered that someone had smashed a hole in the front compartment of the starboard float. The airplane had complete radio equipment, which all had to be removed and completely dried out. Also the engine had to be disassembled and dried out. No one was hurt, but it surely slowed things down.

I have, for as long as I can remember, been very interested in Aviation. I built models from the time I was 10 years old. When Plexiglas was developed, the Army aircraft were the first to use it. I was able to get some thin (1/16") stock and I made a heat-formed windshield for a 36" wingspan Aeronca model I had built.

I took it to work with me and showed it to engineer Gene Woods. He said "I bet we could make one full-size for light aircraft." We were successful and at one time Heath was the largest supplier of one-piece Plexiglas windshields for light aircraft.

It became my job to make drawings from the aircraft for a form of white pine, which a pattern maker would build. It was covered with felt. A large piece of Plexiglas was heated in an oven till it was flexible. Then several people would use clips and stretch the hot Plexiglas over the form. As the plastic cooled, it would keep the shape of the form. Then it was trimmed to the lines marked on the block to fit the aircraft it was made for.

Another interesting project was the 1/5 scale floats for the L-5 aircraft. I was assigned a helper and set up space in the shop to build them. When they were complete, a bag of lead shot placed on them simulated the weight of the L-5 aircraft. They were put in the swimming pool at the YMCA and towed to study the performance and buoyancy. It showed full-sized floats would be OK.

The experience I had at Heath Company in the early years was extremely valuable to me. Then I worked as a draftsman in the tools and operations department at Whirlpool as the war ended. After that I went to a large lumber yard as an Architectual Draftsman. That job only lasted 2 months, as the man who had the job before came home from the Navy and got his old job back.

Then I got a job with a new company from Chicago, the Wilmak Corp., a Radio Products manufacturer. I was their draftsman, pattern maker and model maker. That company soon went broke. All this time I had my home woodworking shop, so when I was unable to find a job, I decided to go into business for myself.

One of the first big jobs I got was making wood bases for record players made by VM Corporation. After making 1500 units (with my wife's help), we lost out to competition from bigger manufacturers. By this time the Heath Company was beginning to make Radio kits (Heathkits). I went to see Howard Anthony and asked him if he had a need for wood cabinets which I could make. He said he had plans in the works, but it would be a while, and if I would like to work in the warehouse for a month or so, I could. I said I sure would. The work was unloading and stocking surplus electronic parts. It was very interesting.

My first order was for 100 cabinets for a crystal radio for kids to assemble. It turned out to be a hot item, for my wife and I made about 1000 units. Then the tube checker came along. We had to have help to provide about 200 cases a week. We hired from 4 to 7 people (my wife and 3 sons also) as other kits were developed: Resistance decade, Condenser decade, FM Tuner, Bridge, etc.

Then Helen Anthony sold the company to Daystrom. After a short time they decided to make all cabinets out of metal and this, of course, cut us out. I now have a one-man Custom Woodworking shop with all the machinery and tools that were accumulated while supplying Heath company. It has been my good fortune to have known Howard and Helen Anthony as part of the fine association with the Heath Company.

Chas. P. Christy, 1987

Remembrances II

When I joined Heath in 1943, Howard lived in a cottage near Lake Michigan Beach, north of town. Then he moved to a neat small house on Seneca Road. Later he moved to a large house on US33, north of town. One time he had a brush fire there and half the company went out to help put it out. Then he got intrigued by Frank Lloyd Wright and got him to design a house for him. It was built on Miami road, close to where he lived before. It drove the builder crazy, for the right angles were few. Most were 30 and 60 degrees. It was a lovely place, inside and out.

Those moves reflect his fortunes. As I heard tell, the first few years of Heath ownership were hard times. The depression was on in full force in 1935, when he bought it. There were times when he was so short of funds that they faced starvation. For a couple of weeks they stole or borrowed canned fruit, stored by Michigan Fruit Canners in the basement of the Baker-Vawter building, so they wouldn't starve to death. When he finally got the payments from the airport operators for the tailwheels he had sold them, he repaid his debt promptly. But his experience with poverty was evident in his relationship with his employees the rest of his life. And they all admired and respected him for his humanity.

At Christmas time in '43 he rented the Aircastle restaurant for the company party and there was food, drink and presents for all employees and their spouses. A year or two later he sketched up a design for a kid's wagon and got a number of us to work after hours to build them for the employee's kids. The traditional wagons were unavailable in the stores because of the war. But he had been optimistic about how long it would take, and it ended up that the experts built them during working hours the following week. They were real fine wagons, with Heath ball-bearing tailwheels! Our son enjoyed his red wagon for many years.

One of the fringe benefits was typical for the Anthonys. Shortly after I joined, Blue Cross was the up-and-coming Health Care plan and Heath signed up as a company for all of us, provided we would sign up for Blue Shield. The fee for that would be deducted from our weekly paychecks. Typical of Helen, she found the added work of figuring the checks was not worth the savings, so by the second check Heath paid for both Blue Cross and Blue Shield!

Howard backed the United Way, also just appearing in this area, by getting the whole company to put in a half day on Saturday without pay, and the earnings went to the fund drive. We all went outside in 3 groups and had our pictures taken. It gave us good publicity in the community! Howard was real proud of the community and often spoke of the value of making things here and selling them all over the country. That enriched the local area. It took us years to make as large a contribution again, for the half day work idea was not repeated.

Howard Anthony had two great loves, beside his wife Helen. One was flying and airplanes. The other was Electronics, which was then called Radio. I don't know when he learned to fly, but he was a good private pilot when I met him. His skills in the field of radio I heard about as follows. After finishing high school in Dowagiac, it seems he spent some time at Hillsdale College, but I don't think he graduated. If he did, he never mentioned it. He said that in Dowagiac he lived on the wrong side of the tracks. It seems possible that he didn't finish college for economic reasons. But he had an interest in radio and probably built one of the famous Meissner radio kits. Seems he set up a little production line for a radio receiver in his parent's garage and some of the neighbor ladies did the building. Possibly zoning ordinances brought that venture to an end.

His experience with building and servicing radios made him value the test equipment needed for that. During the war he spent some of his spare time building an oscilloscope, one of the newest devices, at that time, for seeing what goes on in electronic circuits. They were both scarce and costly. He enjoyed trying the newest circuits described in Radio News and other magazines.

One day a fellow named Don St.Clair from Chicago stopped in and asked for scrap windshield plastic material. We asked what he had in mind and he told he wanted to make chassis for a Vacuum Tube Voltmeter or VTVM he had designed. We offered to make it out of metal, and he liked that better! Then we agreed to make the cabinets too, and even assemble the instruments for him. He had the parts on order, but they came COD and soon he was out of money. Then Howard bailed him out and we ran the assembly line during the week. On the weekend Don came to town and did the calibrating with precision standards. Howard sold the VTVM's to Bill Acker at Radio Distributing in So.Bend until his expenses had been recouped, and then Don got to sell the rest of them. They were fine instruments for their day, and Howard learned a lot from that experience. Later Don talked Wilmak Corp. into building more of them, for we had enough of that fun.

During those war years I designed the HBR-4 and HBR-5 battery-powered receivers for both the Beacon (200-400 kHz) and the Broadcast band. We made a number of them. There were other designs worked out too, but I don't believe they got beyond the prototype stage. For the receivers we ran a small 4-person assembly line. My instructions were step-by-step on file cards and that seemed to work out well. One problem we ran into was excess heat on the tube socket pins, which let the solder run into the contact. Then a tube couldn't be plugged in and repair was needed. We fixed that quickly, by letting the person who had let the solder run down do the repair! A nasty job, but after that it seldom happened again!

Bjorn Heyning, 1986

Joining Heath

For many years, I had operated a radio sales and service shop on Pipestone Street, in the old Elks building in Benton Harbor. Howard Anthony used to drop in occasionally and shoot the breeze about service problems, customer problems, and so on. His interest was genuine since, for many years, he was in the radio repair business in the Dowagiac area. Then, after buying the Heath Company, in bankruptcy, and moving it to Benton Harbor, he became very active in the electronic surplus business.

I first became aware of the Heath Company through ads in Popular Science and Popular Mechanics magazines, small one or two inch column ads, and the idea of an electronics surplus company in the Benton Harbor area was appealing but, offhand, I felt it was a pretty temporary arrangement. However, when kit-form oscilloscopes and vacuum tube voltmeters made their appearance, my interest heightened.

After a call to Howard, I received a phone call from a man named Curly Knapp, who scheduled an interview appointment for me. I arrived at the appointed time (this was at the old Heath plant on Territorial), and the receptionist called Curly's office and announced my presence. Suddenly the door to the reception area burst open, and out came a friendly, rolly-polly type of man, who announced that he would not be able to see me - he had a dental appointment - and would I please come back. Well, I don't mind saying that I didn't feel too enthusiastic about coming back, because my interest was purely out of curiosity. However, I did come back, and it proved to be one of the smartest moves I ever made.

I was hired in October, 1950 as a technical correspondent. No mention of salary, no negotiation; all I was interested in was a trial period to see whether or not I wanted to cast my lot with this company, as opposed to continuing my own business.

My first exposure as to the type and character of the organization came the first week in December when, along with my regular weekly paycheck, I received a second check, the equivalent of two-week's pay; the second check was identified as "Christmas bonus". Feeling that someone in the accounting area had goofed, and didn't realize that I had only been there six weeks, and had done nothing spectacular to warrant a bonus, I took my check into Howard's office and explained the mistake. He very heatedly assured me that it was not a mistake, that this was his practice, and to take my check and enjoy it with my family. This was the first time in my life I ever got a bonus, mostly because all my prior experience had been self-employment.

This practice continued through 1954; Howard Anthony was killed that year, along with five other people in an airplane crash. The company was sold to Daystrom, Inc. Daystrom was an eastern holding company, controlling about ten or twelve varied companies, mostly in the electronics area, except for one company which was in the furniture business.

Daystrom, as a big-time, tightly-run business organization, was appalled at the Christmas bonus practice, and decided to discontinue it. The explanation was that they did not pay bonuses, but that each employee would receive the equivalent of his bonus by a few dollars increase in his weekly paycheck, so that he would not actually suffer any loss.

Gene Fiebich, 1991

Heath Traditions

The Christmas parties at the Heath Company in the early years were legendary. The festivities actually started the morning of Christmas Eve; not much serious work got accomplished that day. People came dressed for a party, and they got one. The plant was decorated, and the areas upstairs, which were the office and kit-packing areas, were reflecting Christmas spirit, and the lower floor, containing the machine shop, shipping area, paint room, and other related activities, also exuded the Christmas spirit. By noon, any effort at work had ceased, various areas had already started their Christmas libation, and arrangements had been made for a Christmas dinner to be served in the production area. This area was cleared of work tables and bins temporarily, long tables installed, and a very fine turkey dinner, with all the trimmings, was served by a catering organization. Additionally, a small four- or five-piece band played, and there was much dancing and merriment throughout the afternoon, including some highly original gift exchanges.

The Christmas parties also ended with the Daystrom regime because they preferred to hold a more formal type of party, and for several years this was a dress-up affair at the Whitcomb Hotel.

Another tradition of the Anthony era was the annual family picnic. These events were usually held at resorts out in the Sister Lakes area and, again, it was a tremendous activity, much looked forward to. There was entertainment for the children - pony rides, races, contests, and a baseball game between the factory and office workers. Huge chicken dinners were served, and Howard particularly liked the gizzards. One year he made the mistake of going into the kitchen to try to get some gizzards ahead of the regular meal. He was summarily ousted by an irate chef, who didn't know who he was, and didn't care - he just wanted him out of his kitchen. Howard meekly obliged.

The family picnics continued on, and are still part of the Heath activity, but they don't resemble the earlier, family-type picnics. The word family is very appropriate, because that was an essential characteristic of Heath during those early years. Everyone worked as if they owned part of the business. They were closely knit, they helped each other whenever problems developed, and worked extra hours without any difficulty - a most unusual situation.

Gene Fiebich, 1991

Heathkit Heaven

Every amateur radio operator in the world can describe the excruciating excitement of their first radio contact. They can remember the special tingle when they heard their call letters in reply to their first, tentative CQ. That's the thrill I got when I heard that I had been accepted as a new design engineer in the Amateur Radio Department at Heath Company.

The Heath Company of the 1960's and 70's was a dream come true for me. I had just graduated from engineering school, in debt from supporting my wife and child, and eager to begin designing Heathkits. I had received other job offers, mostly from defense contractors. Bullpen engineering had little appeal to me even though entry level pay was higher. Heath Company not only had everything I was looking for, but they promoted a design approach that very few companies of any size would ever consider.

Heath had started small. Employees were given lots of responsibilities. Most workers wore many hats simply because the company couldn't afford large support staffs. This concept was applied to product design, too. When Gene Fiebich became the Vice-President of Engineering, he vigorously encouraged the philosophy of "one engineer, one product". Such an idea was rejected by many companies who believed narrow specialization was a more efficient design method. Gene recognized the terrific appeal full product responsibility had to engineers. For any engineer who wanted to take on the responsibility there was no more exciting challenge.

Imagine being handed a sheet of paper full of specifications, a target selling price, a desired completion date, and perhaps a stylist's drawing of what the new kit was to look like. Although a significant amount of work and planning had already gone into the project at this point, it was the beginning of the design engineer's commitment to produce what that paper described. A new Heathkit had been conceived. Now it was up to the engineer to turn those words and specifications and goals into hardware. It would take many months to deliver his new creation, and many more months for the rest of Heath Company's resources to finally produce that first box full of parts coming off the kit-pack line.

The kit-development process was always a fascinating procedure. It never stayed the same. New techniques were tried and implemented where appropriate. Improvements and refinements were tested as personnel, components, technology, and market forces changed.

Kit ideas came from everywhere. Heathkit builders were a constant source of product requests. New products developed by others were often likely candidates for Heath's kitform considerations. Heath employee's personal interests often generated product suggestions. The marketing department was seldom at a loss for what to do next. In fact, product plans were projected in three and five year time frames which were updated every six months. Proposed product plans were discussed, approved or disapproved weekly. As engineering time became available, these plans were assigned to an engineer.

Not all engineers were created equal, so judgment and future product needs usually dictated who worked on what. Less experienced designers were usually given smaller projects; more complex kits might require two or three people. No matter how simple or how complex a kit might be, basically the same development process was followed.

The design engineer had a lot of freedom in the way he approached his task. One day he might work on the basic physical layout of the product. The next might find him sketching circuits for an amplifier stage or calculating component values for a filter section. Technicians became involved with sheet metal layout, printed circuit design, prototype assembly, and wiring harness construction. Hundreds of electronic parts, sheet metal parts, fasteners, and wires were slowly fashioned into a unit. Prototypes were seldom pretty to look at, but they became the foundation for testing the product's performance. Testing, fine tuning, alternate circuitry, a myriad of activities all helped shape the product's performance.

The design engineer usually built a number of prototype units, each one getting closer and closer to the final model. Ultimate performance was the engineer's goal but the over-riding consideration was: can it be assembled by any kit builder and will it work as well as the prototype?

Some complex products were initially rejected as kitform candidates because most people didn't have access to elaborate testing or alignment instruments. Heath engineers met this challenge. They cleverly devised low-cost but effective testing methods, sometimes using built-in circuitry. A receiver's signal strength meter became a voltmeter or an ohmmeter with the addition of a few parts. Heath TV kits included a small VOM for initial alignment and for future work on the unit. As kits became more complex and specifications approached state-of-the-art levels, ordinary alignment procedures couldn't be used. Designers then specified pre-aligned subassemblies. This assured performance while still leaving plenty of assembly enjoyment for the builder.

As the engineer's design was approaching completion, many other departments were getting involved with this new creation. Draftsmen prepared engineering drawings of the structural components. New electronic parts were specified. Buyers in the Purchasing Department began planning for sourcing production quantities of parts. Tooling contracts were let. In the Manual Department, a writer was selected to put his skills into play with as much consideration and dedication as the design engineer had.

The Heathkit manual was a marvel of excellence. Without a doubt, these step-by-step instructions with their three-dimensional illustrations were the epitome of what an assembly manual should be. Without its writers and illustrators, Heath would never have reached its status as the most successful electronic kit company in the world.

For large complex kits, manual preparation often took as long as the design of the product. The first prototype manual was called a preproof manual. It was used for the initial proof build of the new kit. Usually the manual writer, the design engineer and the chief engineer would each build a preproof. As they followed the assembly instructions, they red-penciled their way through every page and every illustration, making suggestions and comments.

While the author was incorporating these changes to his manual, the design engineer was measuring and testing the product's performance to confirm its ability to meet all specifications. Any problem areas had to be resolved in preparation for the proofbuilding phase.

Proofbuilding a new kit was a crucial step in its development. Here, Heath employees representing a wide range of assembly and technical skills were selected to test the buildability of the product and the accuracy, clarity and utility of the manual. Different skill levels were sought in a dozen or more proofbuilders. At least two persons were selected who had never built a kit before. If these novices were successful, we all felt better.

All proofbuilders attended a critique session, where the manual was reviewed page by page. Each builder was encouraged to describe any assembly problems. All had their input. An error-free, easy-to-follow manual was usually the result of the proofbuild process.

Now, we waited for those hundreds and thousands of parts to arrive. Finally the day arrived when the kit pack line was set up to begin production of the newest Heathkit. But the process had one last step. As a final check on the efforts of the entire company, a kit was unpacked, all parts counted against the manual's parts list, and the unit assembled. Again, were all specifications met? Was the manual error-free? If so, the green light was given for the production crew to pack 'em up!

Those were exciting days for a Heathkit engineer. To see your creation advertised in the catalog, in magazines, and talked about on the air (if it were an amateur radio product) was heady stuff. I count that period of my life as one of the best, rewarding in what I learned and was able to contribute to one of the most unique companies that ever existed. That was due in large part to the early leadership and hard work of so many dedicated Heath employees. It was, indeed, a privilege to have been associated with so many dedicated people who felt that Heath Company was a bit of Heaven on Earth.

Larry Houghton, K8ZVF, 1991

Remembrances III

Howard Anthony was a young man, unhampered by prior knowledge of why something would not work. Thus he could do unconventional things, or apply old, basic principles to new needs. Once he bounced a clever idea off me and I gave it some thought. Then, to my regret, I had to tell him that it didn't work on paper. He agreed that my calculations were convincing and felt as sad as I did. But next morning he brought a working breadboard and said that while I was correct, he found it did work, and we'd do it that way! And we did.

It seems the first floats we made were constructed out of plywood and they worked well. But in spite of the protective coverings of paint and such, they were subject to considerable wear from sand on the beach. Then he found some plywood with a canvas-based outer layer. It was much more abrasion-resistant. So the floats I saw all had that material on the lower hull. Then he heard of the potential benefits of some Fiberglas-impregnated plastic, which was light, strong like steel, and very much more abrasion-resistant. That art was still in its infancy, and he hired an expert consultant, Charlie Skogland, to develop a process that would suit our needs. After a few months we had the method, so a female mold and an oven were built. The floats worked fine, I think, but by then there was no market for them.

Undismayed, Howard sought for other applications for Fiberglas construction and thought of a dinghy for yachts. He got the famous John Alden to design one for Heath. Soon we had one built and the dinghy did all it was expected to do. There were test rides on the St.Joseph river with a small outboard motor on it and seven people aboard! Not bad for a 7-foot dinghy! But it had some disadvantages. It was round-bottomed and had no keel. That means you could row it easily, but if you stopped rowing it would turn and go sideways, backwards and on till it headed straight again! And if you stepped into it a bit off-center, it would dump you into the water in a split second! Fine dinghy, but not much of a boat for fishing or other locally popular purposes. Cliff Edwards designed a nice label for it with the name "HeathCraft by John Alden", but still it didn't sell.

Soon Howard had another mold made for a standard flat-bottomed fishing boat. It was stable and we sold several of them. But many came back, for the Fiberglas developed some porous spots and water would seep through slowly. I later borrowed one of the dinghies and had it out at Paw Paw Lake for many years. I enjoyed it, in spite of its drawbacks. Little did Howard dream that only a few years later a large share of the boats built in the world would be made of Fiberglas! Or maybe he did have that dream, and perhaps he was one of the vital links in such development.

Even though it didn't become the success for Heath he envisioned, he certainly enjoyed the striving for a better, lower cost way to make floats, dinghies and small boats for people.

Many of the early light planes, particularly those used for training new pilots, had open cockpits. It was difficult for the instructor to talk to the student, with propeller, engine and wind noise blanking any voice. Here Howard saw a need and an easy low-cost way to help. He had gut-less earphones modified with stub tubing in place of the wires, connected the stubs with rubber hose to a metal Y junction and then a single hose led to a rubber-molded mouthpiece. With this "Airphone" the instructor could speak to the student and be fully understood, in spite of the ambient noise. A beautiful, simple solution!

One of the key men in '43 was Jimmy Pakela, our young, expert machinist. Howard counted on him for many things. But when Howard had an assistant for a little while, the assistant gave Jimmy an order, which was not followed promptly for some reason. So Jimmy was fired. Funny, but Howard hired Jimmy right back, and soon after that the assistant was gone. Maybe Howard took firing rather lightly, for his faithful helper Ralph Lhotka was fired regularly with Howard's angry shout of "Lhotka, you're fired!" and Ralph would smile and say "Yes, boss" and show up next morning as usual anyway. He knew Howard's anger wouldn't last.

One time, the story goes, Howard needed something from Indianapolis in a hurry, so he sent Ralph down in one of his planes. The weather worsened on the way down, and Ralph could not return, for he had no instrument rating. Howard had Ralph return by bus and sent our Test Pilot, Pappy Dillon, down instead. By then even he was not permitted to take off! It took several days before the plane and pilot got back. It sure irritated Howard!

One of the earliest products was the Inspection Cover. It was a dished aluminum disk, with two formed spring steel strips riveted onto the back, and the plastic rings to brace the hole in the airplane fabric. A ring was glued or doped to both the out- and inside of the fabric and then the center was cut out. This hole gave access to control-cable pulleys and such. The cover was then snapped into place to close the hole. We had a punch press to stamp the cover, the springs and the rings. Foot-operated presses were used for the riveting. It looked to me like we sold many.

For the radio chassis we used Whitney hand-powered punches, a foot-powered shear and a hand-operated brake. We also had a spotwelder, made by Peer Inc.(now Teledyne-Peer), a local outfit. Howard dealt locally, whenever possible, in support of local business.

For the float construction there were the usual woodworking tools and many electric hand tools for drilling and for driving the great number of woodscrews holding the floats together. They broke down often enough to keep Al Tabor in Riverside busy rewinding and fixing the shorts and broken-down insulation.

Bjorn Heyning, 1986

Growing Pains

During the early period of Heathkit's explosive growth, the small company enjoyed many competitive advantages. With growth some of the advantages were left by the wayside. Simplicity of operation was replaced by complexity. All in good business format, of course. This was particularly true after Howard Anthony's death in 1954, when ownership of the company was acquired by outside corporate organizations. Change was inevitable. Many changes were for product refinement, at the expense of increased cost. Other changes were in personnel and organization. The employee roster grew from less than 50 people in 1950 to over 1600 in 1986.

A Few of the Highlights

As a very small company, many formal operational procedures were dispensed with. For example, the machine shop made production metal part runs from engineer-developed templates. Dimensioned drawings and specifications for Production and Purchasing were non-existent. The company was moving ahead too swiftly to wait for specification development. Surprisingly, the error or rework level was low.

Another example, the fledgling purchasing group was expected to buy production material based upon a physical sample of a coil or transformer or whatever. It was usually a vendor sample, which, upon approval for kit-use by the engineer, became a standard. Of course problems developed in multiple sourcing and repeat ordering. The standard was gone!

This growing need led to the formation of a strong specification group under the very capable and experienced direction of Ray Freridge. Now vendor samples were subjected to appropriate tests before approval for production use. Components were defined by category, part number, and electrical and mechanical specifications. Then all pertinent information was transferred to a dimensioned drawing. Now accurate data could be reproduced in as many copies as required, and Heath retained its standard. There is no question of the validity of this program, but, like so many others, it did add cost to the product.

In the early days a formal Quality Control dept. was unheard of. Conscientious employees were also informal QC inspectors. Women kit packers noted if a tube socket pin was missing, or a transformer wire was pulled out, and made any number of visual checks on parts.

Of course no electrical inspections were possible. With the advent of the specification program, information on incoming production material was available, but there was no one in the Receiving dept. qualified to make tests.

Technical people were hired, and this was the beginning of a Quality Control group. Over the years QC responsibility expanded from checking incoming material to many other phases of company operation. Here again, there was no question of the desirability or need for a QC group, but it did add cost to the product.

Similar refinements or expansions occurred in other areas with formation of services or functions not previously existent in a formal sense: Personnel, Marketing, Publications, Value Engineering, computerization of detailed operational data, and on and on. A full and detailed accounting of all of the operational changes, thru growth, could fill a book.

Before leaving the subject of change, a rather controversial one is worth mentioning. As the company and number of employees grew, so did the problems. One of these was inventory shrinkage or pilfering. The revelation of a particularly sobering system of stealing finished goods resulted in the installation of a Pinkerton Security Guard system. This led to a requirement for employees to enter and exit the plant at prescribed areas. Outgoing packages were subject to inspection, lunch boxes - brief cases - etc. No materials - sample or proofbuilt products or finished goods of any kind - could be taken out without a package pass issued by the employee's supervisor. Again this system was based upon sound business practice, but the annual cost exceeded the known value of the shrinkage. Additionally there was a circumstance where a plant guard was involved in shrinkage! Just illustrates that no system is perfect.

It should also be stressed that many very positive changes resulted from sequential ownership of Heath by large and remote corporate structures. Among these were the establishment of plant-wide pension plans, fair and standard job classifications along with periodic employee performance reviews, and a uniform vacation policy recognizing and rewarding longevity of service.

In addition, employees were encouraged and financially assisted in continuing educational programs. A formal patent program which encouraged highly creative Heath people to disclose their ideas for the betterment of the company and for their own personal achievement and recognition. A form of profit-sharing which enabled employees to voluntarily contribute to a personal growth fund, with contributions from the company related to current business performance - another nice nest-egg for retirees in addition to pension and social security income.

These changes are not all-inclusive, but do illustrate the benefit of corporate operating philosophy. Changes and refinements over the years added continual improvement. And, of course, just as the pattern in our great society, not all changes can possibly meet with the full approval of all people.

Change related to growth was also evident in advertising and marketing practices. As a fledgling company Heath depended heavily on Direct-Mail advertising through seasonal flyers and an annual catalog. This was augmented by saturation advertising of 12 to 16 continuing advertising pages in two popular trade magazines. In fact, many customers assumed that Heath owned the magazines.

It was logical that further exposure of Heath products in some National trade show be tried. This met with overwhelming success. The Heath exhibit area was constantly packed with eager Heathkit fans. A typical example, in the early 50's, was a Heath display at the National Electronics Conference held at the old Edgewater Beach Hotel in Chicago. My attendance at the show as a visitor was quickly changed, as Cliff Edwards begged me to stay and help staff the display area. A casual afternoon visit was changed into a three day stay! Cliff arranged for hotel room, tooth brush, extra shirts, socks, underwear, etc.

Another prime example: Encouraged by the Chicago experience, plans were made to exhibit Heathkits at the prestigious International Annual Institute of Radio Engineers, now IEEE, show in New York City. The results were unbelievable. A small 10' exhibition booth was solidly jammed with interested visitors, overflowing into adjacent booths and solidly plugging the aisleway ... all day long! The loyalty of enthusiastic Heathkit fans was awesome and humbling.

This loyalty characteristic can best be illustrated by the following incident: In the 60's Heath released State of the Art TV in kit form, starting with B/W and progressing into a series of Color TV's. The acceptance of these products was so strong that it was difficult to keep up with incoming orders. At times an allocation system was used. This activity finally drew the attention of Consumer Reports magazine of Mt.Vernon, NY. The Heathkit Color TV was reviewed and results published. The review was grudgingly favorable, but the product was downgraded on the basis of a debatable technical point in the IF circuitry. My immediate reaction was to take the TV engineering team, Don Rupley and Roger Brockway, to Mt.Vernon and meet with the technical group responsible for the report. The TV sample was checked, design philosophy explained, viability of relative performance characteristics discussed - all to no avail. After hours of discussion and repeated visits, their position remained unchanged. However, Consumer Reports was totally unprepared for the avalanche of protesting letters from loyal Heathkit TV builders. In fact, during a succeeding visit I was told that they had never experienced such an outpouring demonstration of product loyalty and it was necessary for them to print a form letter response to the incoming mail!

Gene Fiebich, 1986

You took us too literally...

[An example of the need to consider the novice builder when writing manuals] was the use of the term "spaghetti" for the piece of insulation slipped over the bare wires. It was widely used in the construction articles in magazines and everyone knew what was meant by it - we thought. Until a Radio kit (maybe an AR-3 or a BR-2) was returned for service. As I recall, it smelled terrible and the technician found a dead mouse inside! The cause was that the kit builder had used macaroni, softened in water, instead of the "spaghetti" we called for, around the bare wires. The mouse had crept in and eaten the macaroni and gotten electrocuted in the process, blowing the fuse! From that day on we replaced the term "spaghetti" with "sleeving" in all our manuals!

Bjorn Heyning, mid-80s

Quality control at Heath

Around 1950 we had a visitor and Howard Anthony showed him all around the plant and through the warehouses. When they got back to Howard's office, the guest remarked that one department he had not seen. Howard doubted that, until the guest said they'd missed the Quality Control section. Howard laughed and said we didn't have one. "But don't you realize how much money you'd save if you bought parts to specification and inspected them for meeting the specs?" the visitor said. Howard thought and said "Yes, maybe enough to pay for the inspector."

The background for Howard's view dates back to his wartime work on government contracts. The resident inspectors were nice people who did their task unhurriedly, and had much spare time after they had rejected something for an insignificant failure to meet a specification. Howard was especially upset when he found one of them sound asleep after lunch, doubled up in a carton behind some others, for that was on his money. After the war he steered clear of government contracts and of inspectors of that sort.

And yet our quality was very good. In fact, the few parts we had that required inspection, such as the modified speaker, which we called a sweepmotor, for the TS-4 TV Sweep Generator, formulated the tenet that if a part needed inspection there were 2 possible causes: We didn't pay the right price or we got it from the wrong source. (Those sweepmotors would run fine to start with, but some would get soft in the spider after a few hours and then the disk would flop against the coil above it.)

We prided ourselves in using reputable sources for our parts, with names such as Stancor, Chicago Transformer, Simpson, Mallory and many others, who were widely known and advertised. They knew that we did not inspect them, and that any failures in the field would be returned to them for full credit.

They knew that any failure would reflect on them, rather than on Heath, and that made them apply their QC to the parts they sent to us. Maybe they charged their inspection cost to advertising.

One time, when Magnavox made FP-type electrolytic capacitors under Mallory license, I noticed that there was no Magnavox name on the cans. So I asked the Sales Rep if he was ashamed of what he sold us. He was shocked and said: "Of course not! What makes you think that?" I said there was no name on the caps, and he in turn asked: "May we?" I assured him that we'd found it the most effective and economical form of Quality Control, and would like to have the name on the parts. The next shipment did.

Actually his position was not illogical, for most Radio makers shopped for the lowest-cost parts, and insisted that any source identification be limited to the Industry (RMA) code, so that competition would not know where the best deal was obtained. And that is probably still common for many parts used in electronic (TV, Radio, Computers, etc.) factories.

But that was over 30 years ago, and much has changed since then. The company became part of a Corporate enterprise, and with the insistence of documenting all of our parts (except screws, nuts and tubes) it was easy to give the vendor a print of the part and ask him to quote on that. It was understood that we would inspect what he delivered, and if it didn't meet the spec, it would be returned for full credit. But once we had accepted the shipment, it was ours, no matter what the field failures showed.

In retrospect, maybe there was more to it. Could it be that our faith in the suppliers spurred them to do their best and to take pride in what they delivered to us? Maybe our pride in what we offered our customers, both in parts and in kit concept, from basic design, through the manual, to the prompt shipment and service after sales, was extended both ways: to the vendor of parts and to the kit builder?

Or was it a form of advertising, with good parts reflecting well on the supplier for making them, on Heath for using them, and on the kit builder for installing them successfully?

Since those early years we still make sure that the parts are the best that serve the purpose, at a price that the kit builder can afford. And, probably, there are fewer field failures now than in those early days. Especially when one considers the increased complexity of circuits.

Actually, most of the present day kits would be impossible to produce and market without the Quality Control system now in use at Heath. Added regulations at the federal level, such as FCC, and the Underwriters Laboratory standards would be hard to comply with from day to day without lots of testing and checking.

Only our established QC record let us sell products for Citizens Band and phone hookup in kit form. Some of these were really significant breakthroughs in overcoming bureaucratic regulations, based on assembled products. It was difficult to convince the authorities that our Microwave oven kit would be within their standards, when home-built per instructions in the manual, but we succeeded. It certainly was demanding for the designer to make it so that the end result would be safe. And only thorough QC could protect the customer against faulty parts making it hazardous.

The first computers may have been relatively easy, but their proliferation soon led to tight FCC specs to limit their spurious radiation into radio and TV reception. Good QC kept us in the market.

Progress surely has been made in he last 30-some years.

Bjorn Heyning, mid-80s

Service First

In the early 50's, Howard Anthony used to open the morning mail and sort the incoming orders from the other correspondence. Monday's mail always brought in the largest volume of orders. Whenever a customer complaint letter surfaced, Howard immediately dropped whatever he was doing and took the letter to the person he felt was most appropriate to respond to the customer. Customer service was the number one priority.

As an example, in the vacuum tube voltmeter, the meter movement itself represented over one third of the total kit price of $24.50. If a customer complained that the needle on his meter was sticking, he would immediately receive a replacement movement, without having to first return the defective unit for inspection and adjustment. Although he then had two meters in his possession at one time, returns were generally not a problem. Customers were pretty conscientious - they appreciated the consideration they were given, and responded in kind. This replacement policy applied to any component in any kit. "Service the customer" was the primary rule.

Gene Fiebich, 1991

Three Years as a Heath Consultant

In the fall of 1951, a former associate of mine, Walther Buhl, called me to say that he was living in St. Joseph, Michigan and that he had met and was working with a Howard Anthony, who owned the HEATH Company. Walter said that Howard had asked him if he knew someone who might be able to design a Q-Meter that could be sold in kit form, and for some reason Walter gave him my name. Howard asked Walter to contact me and arrange for me to visit the plant in Benton Harbor. At the time I was in charge of an electronics design section at an aircraft company in Akron, Ohio and living on the outskirts of Cleveland. The 70 mile round-trip every day, plus the fact that I had never tried design consulting before, made me less than anxious to become involved. But Walter insisted that if I talked to Howard, I would do business with him and that I would enjoy working with him. Reluctantly, and with some misgivings, I finally agreed to fly to South Bend one weekend, where someone would meet me and take me to Benton Harbor.

The day was Saturday, November 3, 1951. It was snowing lightly when I boarded the old DC3 in Cleveland. I don't remember who met me in South Bend, but we got to Benton Harbor safely and there I met with Howard. He explained what he wanted in the way of a Q-Meter and said that if the project was successful, there would be more. He mentioned a laboratory signal generator specifically. As Walter had promised, I liked Howard from the start. He was one of the most energetic people I had ever met. We had an informative conversation about the kit business and how it got started. I'll try to relate my recollection of what he said.

First of all he said he had gotten the company through some kind of a tax sale a number of years earlier and had gotten into the electronics business during the war, when he had a contract with the Signal Corps (I think he said). At any rate he enjoyed putting electronics gear together and reasoned that others would also enjoy doing the same. He mentioned that several advisors at the time told him that such a venture was crazy and that he would lose his shirt. The numbers he gave me that day were that the company did a million dollars worth of business the first year of selling kits (1948 I believe), two million the second year, three million the third, and he expected to do roughly five million in 1952. My recollection is that he told me that his favorite kit was the VTVM and that as long as he was around, the company would keep on selling it. I asked him about getting into the TV business and again my recollection is that he wanted no part of it. Howard told me that one of the big advantages of the kit business is that patent laws did not cover kits so you didn't have to worry about infringement suits.

We set up tentative goals for the Q-Meter on a feasibility basis. The frequency range desired was from 160 kcs to 30 Mcs (note megacycles back then) if possible. It should be able to measure Q's up to 250 or 500 within 10% or better, and it should be able to sell for $69.50 or less. I asked how to determine the price, and as I recall, he said just add up the cost of the parts used as if you were to buy them out of the Allied Radio catalog, and that will be 80% of the price we can sell for. He said that he would buy parts in large enough quantities so that there would be enough difference to pay for the relatively small cost of labor to package the parts into a kit for shipment to the customer. As the last item on the agenda he said he would pay me $5.00 per hour for my work and would provide all the parts I needed to fabricate a working model, that would in turn be turned into a marketable kit by experts like Bjorn Heyning. We wound up the day with a short trip through the factory and then I took off for Cleveland and my first venture into kit design.

By November 24, 1951 I submitted my first development report to Howard with an analysis of what might be expected in the way of performance and some suggestions as to what course to follow. Through Walter Buhl, who was in the process of developing the GD-1 Grid Dip Meter for Howard, I was advised to proceed with the construction of two feasibility models of the Q-Meter for which parts were being shipped. After receiving the parts early in December, I spent nearly all my spare time constructing, calibrating and testing the first model. Having no previous guide lines to judge how efficient I was being, I began to worry that the whole project was getting too expensive, so I wrote to Howard explaining where I stood and presented a list of four suggestions for proceeding, including one that expressed my concern that maybe not enough progress had been made for the money invested. Within just a few days I received a reply from Bjorn saying that Howard was pleased with the progress and that he had no intention of scrapping it. I felt better; after all, $5.00 per hour was good money in 1951!

On March 29, 1952 I finished the final report on the project and sent it off to Howard. Several days later I received a letter asking if I could be in Benton Harbor on April 12th to meet with Bjorn Heyning to discuss the layout and fabrication of the kit. Since April 12th was Easter, I asked if we could make it April 19th instead. Howard agreed. At this point in time I felt pretty well burned out, what with driving 70 miles a day through lots of snow and rain to get to and from my regular job and then spending practically all of my spare time on the Q-Meter. I suspected from previous hints that Howard was going to try to talk me into working on the signal generator, but I was prepared: I needed to paint the house in the next few months and wouldn't have time! Little did I know how persuasive Howard could be.

Since I had some equipment I needed to return to HEATH anyway, I asked Howard if it would be all right to drive up from Cleveland. He said OK and so I started out at about 5:30 A.M. and got to the factory at about noon. Howard was waiting for me, but Bjorn was not available that day. I asked Howard how the Q-Meter was doing and he said, "Fine. I want to talk about your next project." I was ready and said, "Howard, I really can't do another project right now. I need to paint my house." Howard didn't even hesitate in answering, "Oh, you don't need to do that. I'll raise your fee to $7.50 an hour and you can hire a painter, while you are doing something you like to do." With that he reached under his desk and pulled out a box of parts he had selected for me to get started on the laboratory signal generator. How can you argue with that kind of logic?

After discussing the goals for the next project, we went out to his home to pick up a piece of equipment for the next project. My main recollection was that it was obviously a Frank Lloyd Wright design and that there were a number of electronic instruments sitting on the floor of the master bedroom. I got home a little after 11 P.M. that night after driving a total of 618 miles! I never saw Howard again, though I talked to him on the phone a number of times and exchanged several letters. My part of what became the LG-1 was completed on July 25, 1952. The painter had also finished.

I did two more projects for Howard, but they never reached the production stage.

On April 16, 1954 I received a surprise call from Howard at my place of employment. He wanted me to help with the analog computer being developed at that time. I told him that that might pose a problem, because the company was involved in the development of analog computers and related equipment. I talked with my boss, with whom I had cleared my previous involvement with HEATH, but this time he ventured that while the company had no interest in electronic instruments, it had a definite interest in computers. I related this information to Howard, who then asked if I would mind talking to George Zindler at least for one afternoon about what I knew about analog computers. I agreed, as long as we stayed clear of any proprietary information. So it came about that I spent a very enjoyable afternoon with George talking about analog computers, just as Howard said I would. I wrote my last letter to Howard on April 26, 1954.

My contacts with HEATH faded away after Howard died, but he is one person I'll always remember fondly.

Carl Helber, 1988

Trade Shows

In the early 50s, Howard decided do some national advertising in terms of trade show participation. In those days, the IRE national convention in New York City was the tremendous event looked forward to by many in the electronics industry. Exhibition space was available in the Grand Central Hall on the north side of the city. This was our initial, big-time exhibition, and Heath had a ten-foot booth with some of the major products on display, a staff of people to operate the booth in shifts, and a good supply of fliers.

We were totally unprepared for what was to happen. We knew we had a strong customer base in the east, and when it became known that Heath was exhibiting, the attendance was phenomenal. The isleway in front of the Heath booth was so packed with eager visitors that traffic was severely impeded. The exhibitors to either side of our booth were inundated with people who entered their areas to peer over the walls into the Heath booth. This became such a problem for them that those companies didn't bother to show up the last few days of the show.

We ran out of fliers about the second day, and a hurry-up shipment was made from our printer in South Bend. The whole experience was phenomenal; we were all very tired, but it was a very successful event for us. This was repeated in subsequent years, with our display area eventually quadrupling to allow room for the expanded instrument line, the new electronic organs, and high fidelity equipment. When we moved into the display building on Columbus Circle, the practice was to rent a separate room on an upper floor to hold audio demonstrations; hi-fi was becoming a very hot area at that time. Ultimately, these exhibits had to be curtailed, largely due to pilferage by the union show workers who tore down the displays. Many crates were found to be empty upon their return to the factory.

The trade shows we attended in future years were all highly successful, with large crowds of Heath enthusiasts, eager to see our new products and relate stories of products they had built.

Gene Fiebich, 1991

A Tip on Calculators

It was in the early 70's that Hewlett-Packard first introduced their line of scientific pocket calculators. Those engineers who were lucky enough to own one were justifiably proud. But some got carried away with its convenience.

The Chief Engineer of the ham group at that time was Harry Meyers, who wore an HP-55 in a leather holster on his hip, ready to draw whenever there was a calculation to be made.

At one of the ham conventions, Atlanta I believe, Harry and several others had just finished a meal at the hotel restaurant when the checks were delivered. We were speechless when, out of habit, Harry reached for his HP-55 to calculate the appropriate tip.

"Proof-building" Assures Perfection

To ensure the quality and accuracy of the Heath manuals, a "proofing" procedure was established. When a new design was complete, it reached the milestone known as "manual transfer". The product was transferred to the assigned manual writer in the form of an assembled unit, several complete sets of parts, and documentation from the design engineer(s). The documentation would include notes on operation, circuit description, trouble-shooting, and specifications, a schematic and bill of materials, and any other information that might be helpful in writing the manual.

The manual writer would then outline an assembly sequence and build a unit, noting any areas of difficulty. The procedure would be refined and used to build a second unit. The writer would also work with one or more illustrators who would sketch the necessary pictorials in parallel. In the pre-CAD days, the manual department was also responsible for generating a production-quality schematic from the engineer's hand-drawn copy.

When the text and sketches were finished, a preliminary manual was put together and several copies made. This was known as the "pre-proof" manual, and was used by an engineer and manual writer to build units, and trouble-shoot and align them, if appropriate. Both people would make detailed notes concerning any problem areas or errors they found. Sometimes, whole assembly sequences were reorganized to make the kit easier to assemble, but this was seldom necessary due to the experience and intuition of the excellent team of writers. (The writers often came up with ingenious ways of accomplishing difficult tasks, for example, using a piece of the shipping carton as a spacer to properly position something being soldered.)

The proofbuild cycle was handled by the Evaluation Department. About a dozen kits, consisting of production parts and revised manuals, were assembled by manual writer, design engineer, chief engineer, and product line manager, as well as someone in Market ing, Technical Consultation, Evaluation, Service, Quality Control, and others, both to perfect the manual and to familiarize the various departments with the new product. In addition, a "novice" builder was chosen, preferably someone who had never assembled a kit. Each builder also kept track of the time spent on each assembly, the amount of wire and solder left over, and any hardware discrepancies.

The assembled kits and marked-up manuals were returned to the Evaluation Department, which compiled the notes from each manual into a master manual, and performed limited testing on each unit. (Extensive evaluation of performance was accomplished earlier in the design cycle.)

The units were then returned to each builder to take home, with the request that they use them and provide comments on their operation that might result in later improvements.

Proofbuild of a few of the more complex products resulted in them going through an additional "post-proof" step to shake out some unexpected bugs.

Finally, before the first product was shipped, a kit was randomly selected from the line, and used to build a "line-proof". This ensured that all process sheets were correct and followed properly.

Together, these "proof" cycles helped to ensure that the customer received a kit product that was complete, and as easy to build, align, use and maintain as Heath knew how to make it. Most customers will agree that Heath products were hard to beat in these respects.

Terry Perdue, 1991

A Time for Caution

One of the Service Supervisors was Jim Van Duzen, a part-time deputy Sheriff. His group was responsible for repairing consumer electronics products such as electronic clocks.

It seems that one day the receiving department opened a package that contained a second box, with the words "Time Bomb" written on the outside. Being a deputy, Jim took that as a serious problem and promptly notified the local bomb squad! They, in turn, took the package to an open field and "lightly" jossled it with a 12-gauge shotgun blast, destroying the contents. In examining the package after this "test", the squad discovered the remains of a broken clock. The customer, having not been able to get his kit to work, sent it to Heath labeled that way to indicate his frustration.

Needless to say, he must have had an interesting expression on his face when he received a letter from the Sheriff's Department, explaining the demise of his Time Bomb kit.

Ed Mosher, 1991

Welcome to Heath, Earl!

Field Day of 1967 was the greatest of all Field Days for me. That was the year that I became a member of a very special group. The "Mud Lake Michigan Monster Hunting, Crawdad Fishing, and Field Day Society". I had spent time coordinating the three preceding Field Days down at the University of Arkansas in Fayetteville. It felt great to be able to participate in one where preparations were taken care of by someone else, and I could just enjoy the benefits.

What was even more exciting was the fact that this one was sponsored and funded by Heath Company. I was now working for the "World's Largest Manufacturer of Electronic Kits". I had used Heath equipment for a number of years, and was lucky enough to be working, and learning how to design kits, for Heath, after graduating in February.

Mud Lake, Michigan was a Field Day site that had been used by the Heath group for a number of years during the mid 60's. I really don't know how many years it had been used, but it was obvious that the group knew its way around Mud Lake pretty well.

The great people I was working with added to the excitement of my first Heath Field Day, making this a fantastic happening for me. I had no idea at the time that some of the friendships that started that Field Day would help guide my thinking and actions during the coming years. There was no one there who I ever came to dislike. In fact, they were all great guys that totally lived up to the things that have made Ham Radio the great hobby that it is, and Heath the great Ham Radio company that it was through the years. I still remember everyone who was there. Some of the names I don't recall, because of our short acquaintances, but their personalities and their actions I will never forget.

I do remember the names of two men who were to make the greatest impressions on me and have the greatest effect on my life during the years that followed. I am sure that they never knew it, because, like most of us, I have never thanked them for it. It is a shame that we can't occasionally walk up to a friend and say, "Thanks."

I don't know who was in charge of the Mud Lake Field Day that year, but I do remember that Diesel* Doyle Strandlund was the man behind the CW site and Dar Evans was the man in charge of the SSB site. Both Doyle and Dar were engineers in the Ham Engineering Department at Heath. Even though I was never to work directly with either Doyle or Dar, they are always in my mind, and their approach to the way they treated assignments and people have been guiding stars in my developing heavens since that day in June of 1967, when they unknowingly said, "Welcome to Heath, Earl."

* Diesel Doyle was the nickname that Doyle's friends gave him because of the old Diesel Mercedes that he drove and kept patched together with duct tape.

Earl Harris, 1991

Releasing Frustration

The HW-2026 was Heath's first synthesized radio, a high-tech replacement for the HW-202. It was an instant success, but after about a month, reports began to come in, from the more active parts of the country, that the radio was tripping repeaters other than the one selected. An investigation revealed that the transmitter's spurious signals, while usually in spec., were often still strong enough to cause that problem. Simple fixes weren't adequate, so Heath made its first recall of a product. Builders were offered their money back, and an additional compensation, in hopes of getting back as many units as possible. (Many customers who had not had a problem refused to part with theirs.)

Work was quickly started on a replacement, the HW-2036, which used a different synthesis scheme. Evaluation and field testing on the new radio were intense, as management was gun-shy from the first experience. Part of the evaluation was carried out in "Intermod Alley", in the Bay Area of California, with the able help of Howard Nurse, son of company president Dave Nurse.

One of the field-testers, Doug Wise, who lived line-of-sight to the three South Bend TV towers, noticed that if he held the radio just so when tuned to a particular frequency, he could barely make out one of the TV station's audio. Once the mixing combination that allowed this to happen was determined, it was found that other manufacturer's radios could be made to exhibit the same phenomenon under the right conditions. Yet it was still perceived as a problem that had to be fixed. Testing of possible cures took place at Doug's, punctuated by dips in his pool and sweet corn from his garden. This resulted in two tiny L-brackets between the PA and the bottom cover being added. A number of other 'non-problems' also had to be fixed.

After a particularly difficult birth, the HW-2036 finally shipped, and I was glad to see the end of it. At a Christmas party at Bob Fernau's that year, Bob Knapp gave me a gift-wrapped package to open. I was less than delighted to find an HW-2036 inside. But when Bob Fernau handed me an axe and led us all out to a chopping block in his snow-covered driveway, what I had thought to be a cruel joke turned out to be a chance for sweet revenge.

Terry Perdue, 1991

Heath Grows

As the Heath product line grew, the overall sales volume grew accordingly. The product count went from about eight to thirty with an engineering department comprised of Bjorn Heyning and Neal Turner, who reported to me. As the company grew, the engineering department did also as we hired additional employees with specialized skills. For example, as we got into Amateur radio, we hired Roger Mace, just discharged from the Navy, to run that department. Similarly, a young man named Gaylord Russel, who was a rabid audio enthusiast joined the company to take charge of the new audio engineering section.

With the business nearly doubling each year, it became increasingly difficult to keep up with the orders. Obviously, expansion was needed. We were in an old two-story factory building at 305 Territorial, and next door was a large old two-story house that was the center of Benton Harbor's red light activity. Needing space, Howard made a deal with the madam, and bought the house. Prior to having it demolished for the factory addition, he announced, just prior to lunch one day, that he had "bought the whorehouse next door", and if anyone was interested in seeing it, this was the last opportunity. As the entire plant, including the women, paraded through, the director of advertising, Cliff Edwards, stood at the stairway, trying to collect a dollar from each person - he didn't make much money!

The acquisition of the additional property allowed for doubling the area of the plant. The office area was modernized, other areas improved, and efficiency regained. Soon, however, more space was needed, so Howard sublet various buildings, stores, and warehouses in Benton Harbor, until operations were taking place in six locations, in addition to the original Heath plant. Plans to build a modern factory on the east side of Benton Harbor were halted by Howard's death in 1954.

The Anthonys' success in turning the tiny business into a giant, multi-million dollar operation, was such that the thrill of achievement was gone for Helen. She soon began searching for a buyer, and eventually chose Daystrom. During Daystrom's tenure, their only real achievement was to milk the profits from Heath to help support some of their weaker sister companies.

Schlumberger, which acquired Heath in 1962, was more concerned with the business aspects, and less with the social aspects of the operation. Their primary business was in oil exploration and analysis. Although the Schlumberger board of directors visited

Heath frequently, and were intrigued with the Heath operation, and proud of its diversity, they eventually came to regard it as a "hobby shop" that didn't fit with their highly sophisticated instrument reputation, and agreed to sell Heath to Zenith Electronics Corporation.

By 1958, Heath occupied the new plant, designed to their specifications, on Hilltop Road in St. Joseph. By comparison, this was a country club which bore no resemblance to the operation in Benton Harbor, which had become so disjointed and inefficient, that a move was imperative. Although the original 150,000 square feet provided by the new building seemed adequate for any further expansion, additions to the building became necessary several times.

Along with the physical growth of the company itself, many new features were added - new Marketing, Specifications, Evaluation, and other support groups were implemented, a substantial Manual department, with qualified writers, illustrators and photographers, and so forth. In the 70s, the engineering department totaled over 85 people. The Manual department contained another 45 to 50 people. This was a far cry from the small handful of engineers in the early 50s.

Gene Fiebich, 1991

807's in a 104?

Back in the mid-70's, we introduced the SB-104. In order to do so, Joe Shafer, Larry Houghton and I (along with others) went on the road to many different trade shows. We packed all the gear in 16 Samsonite suitcases, including one for the SB-230 (grunt). When the guy who made the foam inserts for the suitcases for us found out that one of our stops was going to be Denver, he provided a cut-out for a six-pack of Coors beer! We delivered. It should also be noted that on all of these trips, Houghton was the treasurer, Shafer found the restaurants, and I did the worrying!

Please Don't Squeeze the Charmaine

Around 1973, there was a young lady who worked in Value Engineering named Charmaine. One day, I went to lunch in the cafeteria, and wound up just behind her in line. In addition to being very pleasant, she happened to be a rather well-endowed young lady, and, like any normal male, I had an engrossing conversation with her as we moved through the line. Imagine my embarrassment when I arrived at the cashier and realized that I had an empty tray! Shafer found out about this (he had a way of doing that) and made like a broadcast station, spreading that information all over the company.

I had to retaliate, so the next time he left town, I arranged for Char to meet his return plane, throw herself at him, and give him a big kiss (a chaperone was involved). Little did I know that his next-door neighbor, who just happened to be a minister, would be at the airport to witness this! Thus began the infamous Shafer vs. Elliott practical jokes.

The other guy's shoes

Back in the early '70s, management decided that we folks in Engineering needed a greater appreciation for the problems faced by manufacturing. The best way, according to our managers, was to run a production line for a day. So, we showed up one Saturday morning to pack kits (TV sets, as I remember). You have never heard so many calls of "STOP THE LINE" in your life, but we did learn about the problems faced by others in the company that day, and that is one of the great things that made Heath a family.

Mike Elliott, 1991

Taiwan Calling...

During the late 1980's, it was common for those in charge of developing or finding new products to visit suppliers in Taiwan, Hong Kong, and Japan. There, we would attend trade shows, visit vendors, and shop for the best value and quality we could find for our customers. During those days, Wayne Wilson, WB8TSO, was spending a lot of his time with Japanese vendors, getting amateur radio equipment, and I was spending a lot of my time in Taiwan, where we had found some excellent sources for meters, power supplies, and function generators.

Wayne didn't particularly like going to Japan, and I confess that it was not really my favorite destination, either. The hotel rooms are tiny, and neither of us enjoyed Japanese food very much. One day over a $40 lunch at a nice spot in Tokyo, he turned to me and said, "Do you know why the people here are so skinny?" I allowed that I didn't. Glumly he said "Because they don't like the food here, either."

The following spring, as I was beginning to make plans for my annual pilgrimage to the fall electronics show in Taipei, I approached Wayne with a pitch for a DXpedition to Taiwan. I assured him that the hotels were more comfortable, that the food was better, and since there were only a half dozen active amateur radio operators in Taiwan, we'd have a ball on the air, and generate a lot of good publicity for the new SB-1000 amplifier he had just brought to market, and for the SB-1400 transceiver and the HK-232 packet/RTTY/AMTOR interface. After a couple of reassurances on the quality of food issue, Wayne agreed, and we started making preparations.

We packed up an amplifier, a spare final tube, the transceiver, and the interface, and headed off to the fall show under the sponsorship of Tim Chen, president of the China Radio Association. We had been warned that it is a major offense in Taiwan to be in possession of a receiver capable of listening outside the standard broadcast bands, and that we had to be very careful on entry to get all the right permissions. These Tim had arranged for in advance.

We landed at Taipei, made our way to baggage claim, where we got all our goodies out of the stack of suitcases and boxes, loaded them on a cart, and headed for customs and immigration. "Look", I told the very nice Chinese officer, "We have all this material that is restricted in your country. We have to have some sort of clearance to bring it in. It is radio equipment." He nodded, checked my papers, and took me to another counter, where another Chinese officer asked if I had any papers describing what I had. I produced the invoice that Annie, our secretary, had made for us. Nowhere on the invoice did it say "radio". It said "transceiver", which might as well have been Greek to this fellow. It said "amplifier", "power supply" and a few other things, but nothing that apparently rang a bell for the officer. He scratched his head, looked for some things in a reference book, and attached a stamp to my passport indicating that I either had to take the stuff back out of the country with me when I left, or pay $135 duty. Then he told me that was all. "Well, that was easier than I had thought", I said, and trundled on out the door to meet Tim and Shane from the radio club.

"Did you clear Security all right?", they asked. "Security?", I said. "All they wanted from me was duty." Shane and Tim exchanged looks that said all was not well here. "Aaaahhh---maybe you had better let us take the equipment to the club station tonight. Right now you could be in a lot of trouble if someone sees this." Thus relieved of our contraband, we headed for the station for initial set-up, and then to the hotel, to get a night's sleep in preparation for the activities of the next day.

Our plan was to do our hamming early in the morning, before our usual round of show and vendor visits, and on the weekend. It did not leave a lot of time, and it meant getting up at 5:30 in order to be down at the club station by 7:00. The first evening, we hooked everything up, but found that the 230 Volt plug was a different style than the 230 Volt socket in the shack. The only remedy was to take a large hammer to the plug, breaking away all the very tough plastic molded around the metal contacts, wrap them in insulating tape, and carefully insert them in the socket with insulated pliers. Not exactly up to code, but that does not seem to be a problem in that part of the world. We fired up on 20 meters, and tuned around. Since it was very early in the morning in most of the U.S., not much was coming in from stateside, but we did hear a YL in the upper southeastern part of the U.S. talking with a friend in Florida. "This is K7OWJ/BV2, Taiwan, break please." Somewhat surprised, she acknowledged us, we exchanged signal reports and then closed down for the night. "I'll bet we made her day." Wayne grinned.

Next morning, bright and early, we were back at the club station. Our Chinese hosts were a little doubtful about working stateside that time of day, but we had reliable propagation forecasts from Bob Brown, NM7M, that said the U.S. should be there in force on the upper HF bands. We tried 10 meters, and quickly had a vigorous pileup going. One fellow in Alaska asked if we could drop down a few kHz. into the Novice/Tech portion of the band, so his daughter could work us. We quickly obliged, and then stayed on that frequency. Most of the big DX guns have BV2 confirmed, and we felt like it was a good idea to give all classes of licensees a shot at the DX. I will say that the equipment performed admirably. We were only running a dipole on 10 meters, but still were getting good, robust reports from the U.S. We set aside part of the time for QRP operation, and were able to confirm contacts with several fellows in the U.S. who were running less than 5 Watts out, and we worked a lot of people running 1000 Watts to a dipole.

One of Wayne's key objectives was to make the first packet contacts ever from Taiwan. He had the HK-232 interface tied in, and had connected to an old -171 portable computer that we had brought along. He got everything working satisfactorily, and quickly made contact with a station in the Philippines, becoming the first packet connect from Taiwan. The first thing the Philippino station wanted to know was whether we would be willing to become a permanent part of the Asian packet network, which, of course, was impossible.

Our Chinese friends were the most gracious of hosts. They put on a dinner for us, presented us with nice wall certificates in Chinese, attesting to our expedition, and generally did everything you can imagine to accommodate our schedules and help things happen. Taiwanese radio regulations require a native operator to be present, and to turn in a copy of the log, so they really had to go out of their way to make all this happen for us.

Contacts with mainland China and with the Soviet Union were strictly forbidden, so we were often in the position of having to announce that we were not able to work those parts of the world. We strictly complied with those rules. However, on the last day, just as we were ready to break down the equipment, I was trying to fish out one last contact, and the pileup was making it a little tough on us. One station kept calling, and I could get his last three letters only. I gave him his signal report, and he gave me mine, and I asked for a repeat on his prefix. By this time, confusion reigned in the shack, because we were saying goodbyes, and Wayne was already packing up the computer and packet interface. You can imagine my chagrin when I finally did get his prefix, and he was a Russian. Fortunately, nobody in the shack but me realized what had happened. Later, he did get a QSL, but with a note explaining how I had accidentally worked him, contrary to the rules.

We didn't rack up the tens of thousands of contacts that some expeditions do, since we were limited to a total of about 8 hours of operation, squeezed in among our business responsibilities, but we did maintain an average of better than one contact per minute, and everybody who sent us a QSL during the first few months we were back got one in reply. Annie made sure of that, and she was the world's most reliable secretary.

Denton Bramwell, 1991

The Thrill of the Build

My first kit was an AR-3 Shortwave Receiver, which I built as a teenager in about 1960. In the years since, I've built at least 300 more kits; even that estimate may be far too low. Yet, even 30 years later, the thrill of opening a new kit never diminished. Even the smell of the packing material still triggered memories of those early kits.

How the packaging people managed to fit so many oddly-shaped parts into such a compact carton without damaging anything always amazed me. Once unpacking was begun, there was no turning back; it just wouldn't all fit back in the box.

Looking back, I think there was something particularly thrilling about building the early products which contained vacuum tubes. The warm glow of the tubes made those products seem to come alive in comparison with products of the current solid-state generation. Nevertheless, the enjoyment of the assembly process didn't diminish much for me.

Regardless of the day of the week that the kit arrived, I usually tried to hold off until the weekend to begin construction. Oh, I'd poke around in the box right away to satisfy my curiosity, but bright and early Saturday morning was the best time to really get started. I can remember times that I didn't even bother to shave or eat breakfast, and worked long into the night, until my eyes refused to focus.

Sometimes, after the last joint was soldered, but before the smoke test, I'd find something else to do for a few hours, as sort of an exercise in restraint. Besides, I knew I'd either be pleased or aggravated when I turned the new gadget on, and I needed to take a break in case it turned out to be the latter.

Usually, the unit aligned just as the manual said it should, and operated perfectly, because I was a careful builder, and Heath always used high quality parts in proven designs. Yet, even though I had been anxious to finish, it was somehow disappointing to be done.

Terry Perdue, 1991

HERO 1 - A Frequent Flyer

Heath's HERO I robot was officially announced and offered for sale in our catalog back in January, 1983. As you might expect, though, a product as exciting and novel as this one couldn't be kept a secret. Months before the official unveiling, we were inundated with requests for demonstrations by various groups. Most of these, we simply couldn't accommodate, but when such requests came from representatives of nationally televised shows such as The Today Show, CBS Morning News and The Merv Griffin Show (to name just a few), we could hardly refuse. The potential free publicity was just too much an opportunity to pass up.

In meeting these requests, HERO and his entourage did quite a bit of traveling in the months before the introduction. Fully functional prototypes were rather scarce at this point. Production had not yet begun, so the few prototypes we had were guarded with a zeal usually accorded the Hope Diamond or some other national treasure. This meant that we could not take a chance on shipping the robot by any conventional means. The ideal way, it was decided, would be to build a special container for HERO and carry it on the plane as a paying passenger. So, a special box was built for HERO and he typically rode strapped into a seat up in the First Class compartment where the larger seats could accommodate his wider-than-human container. His escorts, of course, rode back in Coach.

With the attention on airport security as intense as it is today, you can probably imagine what kind of reception we got when we approached the gate with a box so large it required two people to carry it. Naturally, they wanted to see what was inside.

On each occasion, we dutifully opened the box and went into a long explanation of what this strange contraption was and for what purpose it was intended. (That last one was the most difficult -- "Just what DO you do with it?", was the question we found hardest to answer.)

Anyway, after going through this routine a number of times, someone got a brainstorm. HERO had an electronically synthesized voice and a light sensor, both of which were interfaced to the on-board processor. Why not have a little fun next time we go through airport security?

After developing and downloading a few lines of code into HERO's memory, we were ready. Our hectic schedule was such that we didn't have to wait long before our next plane trip. As always, we approached the security gate with the ominous looks of the security guards focused squarely on our box. By now, we didn't have to be asked to bring the box aside and open it for their inspection. Hefting it onto the inspection table, one of us snapped open the quick-release fasteners, allowing the two halves of the box to come away, revealing HERO in all his glory. This time, as soon as HERO's light detector sensed the change in ambient lighting, the program took control. Without any of us touching him, his head slowly moved from side to side and in his unmistakable voice said, "Boy! It sure does feel great to get out of that God Damned box!"

It's difficult to imagine the reaction of the security guards who, as you probably know, are not selected on the basis of having a sense of humor. Invariably, they would wave us right through amidst tumultuous laughter.

This scene was repeated many times at many airports all over the country. HERO's performance never failed to get us immediate clearance through security. It probably also gave a lot of security guards a story they may be retelling yet today.

James Lytle, 1991

Bob Ashton's Fuse

During the final stages of the VF-2031 proofbuild, I happened to walk by Bob's lab one day, and heard him giggling. Bob had always seemed so serious, and I had not seen this side of him before. Since I was the writer on the VF-2031, I walked in to see what he was giggling about. He was trying to solder a piece of wire to the side of a penny. When I asked him why he was doing that, he said, "Well, I went over to see Bud Simciak to get a new fuse for the radio, and he said to use a penny."

The Smoke Detector That Didn't

When I first came to the Manual Department, I was put in a small office with Curt Pitts and Bob Doty. One day I saw Curt blowing cigarette smoke into one of Bjorn's Delta Smoke Detector kits. He blew so much smoke into the thing that it looked like it was on fire. It never did set off the alarm!

The H-11's Power Supply

The H-11 computer had a so-called "indestructible" power supply. One of the tests was to connect a large wire directly across the output of the supply and turn it on. The supply was supposed to shut down with no damage. When I tried this test, several electrolytics blew paper and smoke all over me and the hallway. When I explained what happened to Harry Archer, the engineer on the project, he said that the test HAS to work. When he tried it, he also blew up his H-11 supply. Fortunately, the power supply was completely redesigned before it shipped.

Randy Kaeding, 1991

Our Memories of Heath

January 1967 saw the beginning of my 17 year voyage through the world of Heathkit. As an engineering technician in the amateur radio and communications department, I worked with a very diverse group of guys. Joe Shafer, then Chief Engineer of the department, kept the group on track. Dar Evans was the 'rock' of the department, and everyone looked to him as mentor. Doyle Strandland was the 'mad scientist' of the group. His prototype of what was to become the SB-100 amateur transceiver was spread out over about 6 square feet of copper-clad circuit board material. As each new circuit was developed, it was added to the maze.

I particularly recall the delayed VOX (voice control) circuit Doyle developed. It annoyed Doyle that existing VOX designs always cut off the first syllable of each word before the transmitter was activated. Doyle spent weeks perfecting a design that would eliminate this problem. His test of the new circuit was to speak the name Don Hoff into the microphone. (Don Hoff was an engineer in another department.) In most VOX systems, that name would be heard as 'on-off'. Unfortunately, although Doyle's design eliminated that problem, it was never used.

Pat Duffner, another talented engineer, became the Chief Engineer when Joe Shafer was promoted. He managed to persuade several of the ham engineers to obtain pilot licenses. I remember that the development of a microwave oven kit during this period kept the department full of gingerbread cakes during testing of the microwave 'stirrer' design.

The engineer who developed some of Heath's successful QRP (low output power) ham products was Bob Fernau. Bob was also an accomplished accordionist. Bob had a habit of 'sending' CQ in Morse Code on anything that would make a noise - tapping his pipe on his desk, or squeaking his chair. (His only responses to those CQs were moans from nearby workers.)

When I started at Heath, I was assigned to assist Ron Hall, who was designing marine radiotelephone products. Ron's products gave Heath a respectable reputation in that market. Thanks to Ron, I was given the opportunity to develop many projects on my own, and to advance up the ranks in engineering and then management.

I remember my co-workers in the ham department. There was Jerry Massay, who became a pilot and moved on after a few years. Jack Keppler saw an opportunity in southern Illinois and moved there. Bill Wheeler developed several products before leaving. Joe Tasketa was the only Italian engineer we ever had. His greatest love was making wine. Then there was Bud Gauker. Bud, Bill Denk and I car pooled for many years. I'll never forget the time that Bud was puzzling over a design while at his bench, and scratched the side of his head, as one might do with a pencil. When the hot soldering iron in his hand met with flesh, he let go of a few words I won't repeat. There were others, but over 20 years have passed since those days, and even the memory of good friends can fade with time.

In reminiscing with another former Heath ham department technician, Bob Mann, shortly after being asked to contribute to this book, we recalled one of many humorous pranks that helped to punctuate our work.

We had a 15-inch loudspeaker in the line that had a metal cover on the back which was just large enough to conceal a small transistor radio, battery, and mercury switch. After making this modification, we secretly replaced it in Parts Department stock.

Later in the day, we returned to the Parts Department, and asked if we could select one of the large speakers for use in Engineering. As Parts Department personnel watched, we inspected several speakers, holding each one up as if to listen to it. You can imagine their surprise when, as one was picked up, a local radio station came through loud and clear!

All of us who were at Heathkit in the 60's and 70's were very lucky to have experienced an era that is now gone. I am sure there are many Heath employees who feel the same way. I recall that during the years I worked in the Quality Assurance Department, our president, Dave Nurse, came by my office on a walk through the entire factory, wishing everyone a good morning between 7:30 and 8:30 each day. You could set your clock by his visit.

My wife, Beth, is another ex-Heath employee. Some memories of her 13-1/2 years at Heath follow.

I started working in the Keypunch department at Heath in January of 1970. We had to enter from the rear of the plant, where a guard was stationed, and punch in and out with time cards.

In those days, catalog requests were of such volume that 90% of the department was involved with keying in that information. The requests came from every state, and from people in all walks of life - students, inmates, actors, politicians, etc. It seemed that everyone knew the Heathkit name.

From Keypunch I transferred into the Schlumberger Accounting Department. They still did all the accounting by hand in ledger books. The financial statements were all done manually and then typed. The Schlumberger stores were growing by leaps and bounds, and new stores were added continually. I believe the store count rose to 50 or so. Of course, with this kind of growth we were eventually computerized.

The General Office area was always kept immaculate. Everything was put away at the end of the day. There was plenty of space for all workers, and everyone worked well together.

I remember when no woman was allowed to wear slacks to work. Eventually, pantsuits were allowed, but everyone had their own idea of what was considered a pantsuit, and before long most any type of apparel was allowed.

No one will forget the days when Kay Rapp ran the cafeteria. Those were the days of free coffee. Many people arrived early to have coffee before work, and some great friendships developed there among people from various departments.

Other fond memories:

The days of the Christmas dances at the Shadowland Ballroom; the bands that entertained, and the good food that was served. The Heath picnics that allowed employees' families to meet each other and enjoy a day together. The cooperation that produced incredible Heath floats for the Blossomtime Parades. The many company-sponsored athletic activities - softball and volleyball teams, golf, bowling, and tennis leagues. The formation of the Heath Management Association, which provided many great motivational speakers for their dinner meetings. This organization kept all of management pulling together for the benefit of the company. The excitement of being selected to proof-build a new kit product, and the pride of having it work when completed. I still have the first radio that I built. The Saturday Salvage Sales that most employees attended at one time or another.

These are just some of the things that I remember. I have great memories of all the friends I made at Heathkit. We wish the best to all of our Heathkit Friends.

Rich and Beth Renken

We're very flattered, but...

In 1979, Heath was at the height of its glory, with $100,000,000 per year in sales, and a real first class engineering department. That was about the time Zenith Electronics bought Heath, in order to get into the computer business, and from then on, Heath sales dwindled by an average of 15% per year. Some say it was because Zenith was cash starved, and routinely stripped resources out of Heath, and others say it was because assembly costs had declined to the point that it was cheaper to make an assembled product than a kit. I suppose no one will ever know for sure, but the net result was that Heath was in trouble.

In the late 80's I was managing both the kit computer line, which was very lucrative, and the test equipment line, which was a real evergreen. We had held our sales decline to about 5% per year in test equipment, and it was obvious that if we were going to grow the line, we had to do something fairly dramatic. I estimated that I could keep the decline at the 5% level, and still put two people to work on a breakthrough project.

Mark Rapaich and Rod Brahman were to be my breakthrough team. Mark is contrary, independent, disrespectful of authority, a superior engineer, and one of my best friends. Rod is quiet, small of stature, and writes wonderful computer code. Together, they took up the task of creating a low-cost, high performance digital storage oscilloscope that we could sell through the catalog and private label for a few other test equipment manufacturers. We almost made it.

We started at square one, and asked the question, "If we had never seen an oscilloscope, given today's technology, what should it be like?" We concluded that the current crop of Tektronix products were user-hostile. The HP offering seemed better, but they had chosen a strategy of using a small collection of pushbuttons to do everything, and that meant that you had to go through two or three steps every time you wanted to change the volts or time-per-division. That seemed a bit awkward.

We set as our goal a pair of oscilloscopes, one with 50 MHz. bandwidth, and the other with 100 MHz. At 20 megasamples-per-second, you avoid a lot of costs, and if you are only looking at repetitive events, 20 megasamples is fine. We also set as our goal that the instruments should "look like and cook like" regular oscilloscopes, and that anyone familiar with a regular oscilloscope should be able to run one with no instruction. Finally, we decided that the response of the machine to changes in set-up should be at least as fast as the HP offering, which was excellent.

We selected a front panel control strategy that involved separate knobs for vertical sensitivity and time base settings (which are by far the most used), and a set of function keys under the CRT that could be used for less frequently set selections, such as trigger coupling, etc.

Mark and Rod dived in with a will, and, in about a year, produced a prototype that would store and replay waveforms, but only at a 3 MHz. bandwidth. The front end portions of the instruments, which would give them their full bandwidth, were up and working, and doing well, but at this point had not been integrated with the rest of the package. Rod had done such an excellent job on the software, that the knobs, buttons, and menus all responded just as we had wished. He was getting more speedy response out of a little 64180 microprocessor than the big guys were getting out of their high powered chips, and that is quite a credit to his skill. The human interface was nice -- everyone agreed that it was the best user interface they had seen.

Unfortunately, the decline of Heath continued, and the project was canceled as the company went through a reorganization. Mark and Rod found themselves out of work, but were quickly picked up by Zenith Data Systems, now a part of Groupe Bull. I transferred to another department, but that department was soon hit by cuts of nearly 33% of the force, and I found myself part of those now recurrent RIFs.

Nearly a year later, I got a call at home from Mark. "We just had the Tektronix rep in, and he showed us their latest. From the front panel, it appears to be exactly the same as our digital storage scope. I even had Rod come in, and he navigated the menus without any difficulty whatsoever. How the heck did they get our human interface?" I confessed that I had no idea.

A couple of calls to Oregon did surface one story, which none of us can confirm. According to that source, Tek had been aware of our project (which we knew) and had been worried about us infringing their patents or relying on Tek-proprietary knowledge. The source said that they had bought off one of our techs, in order to check it out, and had concluded that we had not. I suppose that knowing the project was canceled, they may have concluded there was no use wasting a perfectly good human interface design, and put their electronics behind our menu and front panel. Perhaps someone from Tek who knows the story will see this and come forward with the historical details from their point of view.

Denton Bramwell, 1991

Humorous "Hi Tech" Anecdotes

Working closely with a steadily growing group of highly creative people was bound to result in humorous releases of personal expressions, diversions or tensions. Some of them:

In the mid-50's an engineering program was launched to develop a kit for an Analog Computer. Carl Heald was the project leader. The design was a large desk type unit. It was based upon vacuum tube usage and required several pieces of peripheral equipment. After many months of almost frustrating work in this new field, Carl finally had a prototype ready for demonstration to eager company officers. The product was set up - basic operation explained and then the switch was turned on for performance demonstration. The immediate result was a steady series of rising smoke puffs. Excitedly Carl shut things down and did a fast troubleshooting check. He found nothing wrong and the test was resumed, with more smoke and in greater volume. More checks - no problems - more smoke. Finally the problem was solved by uncontrolled laughter from an adjoining lab area. Chalmer Jones had sabotaged the demonstration by running a length of vinyl tubing through the wall into the back of the computer and then puffing cigarette smoke at appropriate intervals. No reprisals, because Chalmer Jones was the son of T.R. Jones, Chairman of the Board of Daystrom, Inc., Heath's owner at that time.

Another: During development of an electronic depth sounder, it was essential to be able to conduct repeated water immersion tests. This was difficult, for lakes were not readily convenient and were frozen in the winter. So a 20 or 30' masonry horizontal tank was built in the basement of the old engineering building. After it was filled with water, it served the intended purpose admirably, even tho the basement was a dark area. Soon strange "glitches" appeared in the depth sounder readout. Worse, they were variable, no set pattern. Equipment was checked and rechecked. No change. Finally the solution: Some unnamed wag had released a half dozen turtles into the tank. Of course they were the "glitches" as they slowly swam through the sounding beam.

One more: During the Daystrom regime it was decided to release a tape recorder transport deck thru the Heathkit Audio product line. The tape recorder was owned by a Daystrom subsidiary and was never a successful product. Under duress, Heath began working with the sister company, trying to improve the product by problem reduction. After two years of an almost fruitless working relationship, it was decided that the tape deck would be demonstrated for the benefit of a forthcoming Daystrom Board of Directors meeting at the Heath plant. The demonstration was to be held in the Lab area. The engineers involved in this effort were so frustrated by lack of progress, and so disillusioned about the residual problems and chances of success as a product, that unknown to me they took matters into their own hands. Came the big day, the Directors were impressed by all of the "improvements", the switch was turned on and the equipment activated. The results were astounding! The two spinning 7" tape reels were projected about 4 feet skyward and fell, still spinning, and spewing tape all over the bench. Thank goodness the Directors had an unanticipated sense of humor that saved the day for Heath engineering, namely Dave Lewis.

Much more could be written about the early days, the growth and the changes at Heath. However one basic characteristic that deserves high tribute and recognition is the unquestionable loyalty and dedication of a fine group of employees who didn't act like employees. Each did his job in the spirit and attitude of owning the company. Working relationships were beautiful and productivity amazingly high, consistently so. As problems developed, as the result of mistakes, the emphasis was on banding together and getting them resolved. Not on trying to identify or crucify the person or group responsible.

This high level of morale was the direct result of the practices and relationships of Howard and Helen Anthony. In those years, under the Anthony's, Heath was indeed a great family. There were new additions, romances, weddings, births and deaths. A period of fond memories for this writer, who was proud to be part of that era of Heath's amazing growth.

Gene Fiebich, 1986

The Salvage Store

I had worked at Heath for about a week when I heard someone mention a salvage sale. Asking what that was about, I found that every Saturday morning, the small, rundown house across Hilltop Road from the factory opened to employees as a salvage store. It was often not economical for the company to unpack a returned kit to see if any parts had been removed or damaged, or the packaging disturbed in any way, then pack it up again for resale. So employees were given a chance to buy what were usually perfectly good products at about a third of their catalog price - often much less. A list of items to be sold was usually posted on the company bulletin boards on Thursday.

Being an avid experimenter and builder, I was one of the first to arrive the following Saturday. By 8-oclock, when the door was unlocked, there was quite a crowd lined up. Getting there early didn't give me any advantage, however. As I walked in the door, I was told to draw a number from a box which determined my turn to make a purchase.

About 8:15, after having a chance to look around, numbers were called. The first time through the numbers, each person was able to buy a single item. After calling the highest number present, the process repeated, only on this round you could purchase as many items as you wished. To make things even fairer, you could only buy an item during the first round if you hadn't already done so in that calendar month.

I have to admit that I went a little crazy that first week. I was staying in a motel while looking for a house, and suffice to say that I piled up quite a stack of boxes in the corner of my motel room. Not realizing that kits could be returned the following Saturday for a refund if they were found to be incomplete, I threw away a digital multimeter kit I had paid $30 for when I found that the nixie tubes and power transformer were missing.

In later years, the store began to open every other week, and eventually only once a month or two. Yet, during my years at Heath, I saved many, many times what I lost that first week - on kits, parts, and assemblies that I was able to find uses for.

Terry Perdue, 1991

Becoming "One of the Hams at Heath"

Ever since being licensed in 1952, I always thought that it would be kinda "neat" to make ham radio my career. That dream came true for me in 1963. During those early years, Heath was in its "ham radio heyday" and a household name with electronic enthusiasts worldwide. The "ham" group was a close-knit bunch of dedicated, hardworking individuals, and was more like a "family" to us all. We not only worked together, but also spent many "off-work" hours together as families.

Working at Heath during those days was like being a "kid in a candy store". We were encouraged to help ourselves to what we called "our FREE stock..." of components. Many of the successful ham radio kits evolved in our basement workshops.

The Birth of Heath's QRP Line

One day, during lunch, I decided to stay in my lab and "play around" with a low power 15 meter RF amplifier. I "tacked" together a transistor, a few resistors, coils and capacitors. It looked like a "ball of hair". I then connected a signal generator to the input and looked at the output on a scope. It looked pretty good. I then thought, "I wonder what would happen if I hooked it up to the 15 meter beam antenna." Not even remotely realizing what was to follow, I proceeded to work twelve countries during the next hour. I thought to myself, "Hey! This is fun!" I decided to check the power output after lunch, and found it to be a walloping half watt. I thought if I had this much of a thrill with low power, perhaps others might like the experience also. Thus evolved the HW-7, followed by the HW-8 and "Top-of-the-line" HW-9 [low power transceivers]. These rigs are still being used worldwide. It's always a thrill now, to work someone who is using a rig that I designed.

Naturally, over the years, there have been many articles with modifications and improvements to the HW line to improve their performance and operating convenience.

At the present, I am actively involved in QRP CW operation on all HF bands. I find it to be a challenge, and a never-ending thrill to communicate worldwide with 2 Watts or less.

Bob Fernau, W8NDG, 1991

We nearly have an extended stay in Seoul...

You hear people talk about the exotic smells of the orient. Let me tell you, it is a peculiar mixture of exotic spices, two cycle engine exhaust, and dung. That is true for Taipei, and for Seoul, where we had an airline layover that nearly turned out to be a lot longer.

We often took some of the engineering staff with us on our trips to the orient, because they could help us qualify products on the spot, and help with the resolution of safety and performance issues. This made it pretty nice for the managers, because if someone made a performance claim, we could just turn to the engineer and say, "Come back in an hour and tell me if his claims are true." If they were, we would often cut a deal on the spot, and have the purchase order in place within 24 hours. Larry Bailey was excellent in this role.

I think Larry really liked Taiwan. The food is terrific, the people are charming, he had a myriad of things to see, and his main job was to be the engineering authority that ruled on the quality of other people's designs.

On one of these trips, I had to make a side trip to Tainan, so I checked out of the hotel and parked most of my luggage in Larry's room. When I got back, it was Sunday afternoon, and Larry was nowhere to be found. I was starting to feel the need for a shower and clean clothes, but there was nothing to do but leave a message for Larry at the front desk, and go do some sightseeing and eating of my own.

If I had realized that the young ladies at the front desk do not just deliver the note, but also call the recipient and read it to him over the phone, I would have chosen a statement other than "Surrender my underwear or face the consequences." Chinese ladies are very modest. It must have been difficult for her to say that with a straight face and without blushing.

On the way back to the U.S., we had a short layover in Seoul. Preparations were underway for the Olympic games there, and security was astonishing. It is the first time I have ever been searched and had my baggage X-rayed getting off an airplane. We wandered around the terminal, bought a few souvenirs, and chatted with some LDS missionaries on their way home. When it was time to go, we all went through the necessary screening, and then lined up in a stairwell, waiting for the bus to take us out to the plane.

The stairwell was evidently not hooked to the air conditioning system, and the temperature must have been 105 F, with 99% humidity. We waited, and were told that there was trouble getting the cargo door properly latched, and that we should all go back inside.

Another trip through security. Another wait in the stairwell. Finally, a bus! We arrived at the plane, and were at the end of a long line of people waiting to climb the mobile stairs and fly away.

At the base of the stairs there was a person in a military uniform, with a rifle in hand. Larry said, "The folks back home will never believe this.", and whipped out his camera to take a picture. Quickly, about three Americans, military people going home on leave, turned to me and quietly asked, "Is that your friend? If he is, you had better tell him to put it away, or it will be confiscated, and he will go to jail. In Korea, public airports are considered military installations."

The line was long, so the guard had not noticed Larry. I stepped between Larry and the guard's line of view, and sharply said, "Larry, put that away!". Larry had never heard me use that tone of voice, and he just looked at me, puzzled. "Just do it, now.", I said, and Larry complied. Then I was able to go over and quietly explain the situation, and we were able to continue our trip without major incident. The Americans who had warned us assured us that we had just had a very close call that would have taken several days to straighten out.

Denton Bramwell, 1991

Getting Even...Almost

Near the middle of the development of the HW-2036 synthesized 2-meter transceiver, lab tech Bob Folkert was building prototypes for field-testing, and I was checking them out and aligning them. Usually an able tech, on this particular Friday Bob must have had his mind on the weekend, as several units he built were non-functional. After tracing the problems to transistors and diodes in backwards, wrong values, IC pins bent under, etc., I was annoyed, to say the least. When Bob asked if he could field-test a unit over the weekend, I saw a way to get even.

I filled an empty transceiver cabinet with a few lead plates from the HD-1410 keyer to equal the weight of an actual radio. I put a real PA on the back, and controls, switches, and a lighted meter on the front. It appeared to be the real McCoy. But inside there was a low value carbon resistor and a latching relay, so that the radio would begin to smoke shortly after being turned on, and could not be turned off.

I expected that Ron would call to cuss me out sometime over the weekend, but the call never came. Come Monday morning, Ron brought the radio back in, but gave no hint that anything had happened. Could it be that he didn't get a chance to use it? Finally, trying hard to keep a straight face, I asked how he liked the radio. Keeping as straight a face, he told me of several good reports he had received on the air. Puzzled, I thanked him for his testing and took the "radio" back to my lab to inspect. The resistor was intact, so he obviously hadn't turned it on. So I took it back to him and asked him to make a power out check for me. He came back a few minutes later with a long look and said apologetically, "I think I just burned something up".

Weeks later, Bob admitted that my joke had worked just as I had planned. It had smoked up his basement before he finally turned off the supply, and he had replaced the resistor so I wouldn't know. He had sure taken the fun out of that practical joke!

Terry Perdue, 1991

Who shot JR?...

One of the nice things about working at Tektronix was that I could go anywhere in the world, and if a technically cognizant person asked where I worked, I could tell him, and he would immediately recognize the company I worked for. The only other place I ever worked that had that kind of recognition was Heath. There was also no other company that I worked for that held anywhere near the place in my heart as Heath did. So what caused Heath's decline? Many things.

I think there is a lot to be said for the fact that we had a cash-hungry parent company for a long time. There is a basic rule of business that says your maximum growth potential is closely linked to earnings, in the long run. If you lose money, you will be forced to shrink your company, or redeploy your assets in a more profitable way. If you have already done an optimum job on deploying assets, you will shrink your company. If you are profitable, you have a much happier set of decisions to make.

Heath was profitable in the early years under Zenith, but Zenith was mainly interested in using the Heath resource as a springboard into the computer business. After all, Heath was one of the very first companies to offer a home computer. And you can't fault Zenith's business direction: they built a business that was much larger than $1 billion per year out of a company one tenth that size. However, the fact remains that Zenith was strapped for cash, and Heath, along with other groups, sacrificed resources to feed the mother company's need. That surely did not help. Heath's peak year was the year Zenith bought the company.

Surely another factor in Heath's decline was a shift in the fundamental economics of electronics. Back in the late 40's, when Heath started, electronics products ran on vacuum tubes, and contained a lot of labor, since point-to-point wiring was the order of the day. Printed circuit boards, automatic stuffing, and flow soldering changed all that. The labor content of electronics products declined across the board, and a good, well-documented kit was more expensive to produce than a finished product. Of course, many people (me included) build for the joy of it, but saving money is a big item for a lot of people. That put economic pressure on the company.

The information system in the company was abominable. Zenith was in the computer business, for heaven's sake, but was terribly disingenious in their use of the tool to run their business.

You could never trust the reports that told you what your inventory status was, and frequently two reports that were supposed to give you the same numbers would differ by a large percentage. Inputs to the system were done by hand. Forecasts, for example, were handwritten modifications to computer printouts, which were then re-keyed into the computer. Until just before I left, we took orders by hand, and entered them into the computer later. The local appliance maker was light-years ahead of us in the application of modern tools.

Deming is right: 85% of the problem is the system. I must have spent a quarter to a third of my time straightening out problems visited on us by the "system".

One of our sins was not turning down enough good product projects in order to concentrate on excellent projects. It was uncommon to calculate a rate of return on a project, and very uncommon to have a project turned down.

The store chain had been the "baby" of Bill Johnson, president of the company, and as we entered the 80's, the stores were becoming less and less the servant of Heath, and more and more the servant of Zenith computers. In the later years, the stores were becoming openly hostile to Heath, and wanted Heath products out of "their" stores. A compromise was struck, where a limited number of Heath products were to be carried and displayed, and the rest would be conveniently available by a red phone at the counter.

One of the store employees was charged with working up the standard store display, and with putting the products on the list on display. He happened to be one of the most vocal opponents of Heath products.

On an appointed day, Joe Schulte, the president of the store chain, some of us Heath people, the man in charge of the display, Bill Johnson, and some others assembled in the St. Joseph store to review the display. Up until the last minute, the fellow in charge of the display was barking, "That product--out! That one too!" It did no good at all to point out that the products he was dismissing were those that the top officials of the company had agreed should be displayed. In addition, the display was in the darkest, most remote corner of the store, on old, ugly shelving, and without any signs at all. The store people were having it totally their way, until Bill strode into the meeting, a few minutes late. He took one look at the set-up, and said, "This is not what we agreed to." Then he put me in charge of redoing the display. I could scarcely contain my glee.

We sourced some nice, modern, clear plastic shelves for the slatwall, worked up signage, put out prices on the products, posted a list of other products available on the red phone, stocked the shelves, put in lights, and made a really spiffy-looking display. It was quite a work of art, and, at $700 per copy, plus the inventory to stock it, not too expensive. Bill said he liked it. At that point, I knew he must have a spark of genius.

Unfortunately, our joy was short-lived. Heath had financial responsibility for supplying the lights and shelving for all the stores, and these became a casualty of budget cuts. Within a few months, Heath products were out of the stores entirely. Loss of this channel was disastrous.

Most of our kit computer sales had been through the stores, and kit computers were our biggest money maker. Now we had to rely on the catalog, on our educational sales reps, and on the sale of lighting and security equipment through home centers and hardware stores.

The catalog crew always maintained that they could get the catalog into homes for about $.55 a copy, including postage, and that it was therefore silly to spend advertising money in any other way. Undoubtedly, the cost of producing one more catalog was $.55, but if you took into account the total cost of running the catalog and order entry departments, it amounted to a good third of catalog revenues. Nobody can stand that kind of cost of distribution for long. We either had to cut our cost of distribution in half, or double the number of products sold through the catalog. Neither was in the cards.

So what put Heath into that long downward spiral? Many people who loved the company will put forth ideas different from mine, but I think the issues were a cash-starved parent company and the accompanying loss of resources, changing times, an absurd information system, lack of financial acumen in the selection of development projects, loss of a critical channel of distribution, and distribution costs that were completely out of line. It was certainly not the fault of the core group of dedicated Heath people who struggled to surmount the unsurmountable.

Denton Bramwell, 1991

Will History Repeat at Heath?

History repeats itself, and at Heath that is clearly visible. The enthusiasm of Ed Heath for the fun of flying dated back to his first plane, which he built and flew in 1909. He wanted to share that fun with others, and after many years he found that war surplus made it financially easier to help others to fly. But it wasn't until 1926 that the Heath Parasol bought flying within reach of the multitudes, especially in kit form. Ed Heath was killed in a flying accident in February 1931. The investors, who took it over, renamed it International Aircraft Corporation. It lacked the spark of Ed Heath and lost the struggle for survival during the Depression. After a move to Niles, MI, it went broke in 1935.

Howard E. Anthony, an ardent flyer and radio experimenter, bought the remains and moved it to Benton Harbor, MI in 1935. There he restarted the business on a small scale with tailwheels and other useful accessories for small planes. After some 5 years he added a small radio receiver for light planes to his line of products. His innovative approach helped the company grow with war contracts, but by 1946 that ended and the business nearly folded.

Again the company dealt in war surplus and Howard Anthony had a chance to apply it to his love for radio by making useful service instruments available at low cost to many others in kit form. Soon the Heathkit name for simple, reliable, low-cost test equipment was as famous as the Parasol had been.

Disaster struck as Howard Anthony was killed in a flying accident in July, 1954. This time the company survived, for the new owners, Daystrom, Inc., retained the key personnel. It had been one of the conditions of their take-over. But the goals of the company were shifted from sharing the joy of tinkering with radio (now called electronics) with the multitudes at low cost, to making ever more profits from the kit builder market. The approach also changed from Engineering to Marketing.

Thus the company continued to grow, as the emphasis slowly shifted from kit-builder-type low-cost kits to more refined and costly general consumer type products. The Hi-Fi and TV lines peaked in the 1960's. Then offshore production brought imports to market here for less than our Bills of Material. That made the kit market start to wane. The take-over of Daystrom (and thus Heath) by schlumberger in the early 60's provided greater independence, but did not change the shift from low-cost mailorder distribution to company stores, or from economy-priced to costly state-of-the-art products.

In 1979, Schlumberger sold the Heath operation to Zenith, who wanted it primarily for the new computer line. These products started as kits, but soon a major share was sold assembled, under the Zenith Data Systems label. Meanwhile, the key staff from the Howard Anthony days dwindled through retirement. Addition of a line of low-cost, simple experimenter-type kits may be the last fling at retaining the kit idea.

And what does the future hold for Heath and its kit concept, which made Heathkit a world-wide household word? My crystal ball shows the following scenario:

Continued losses by Zenith in its main field of consumer electronics (TV) and fierce competition in the computer field may lead to phasing out the kit line and the Heath name. It seems unlikely that Zenith would fold, but the end of Heathkits appears near.

Maybe a daring, young and capable genius will see a future in the Heathkit idea, and buy the name and goodwill of Heath for a small figure, With that, and some novel ideas in an as yet undeveloped line, he might share his love for that line with the multitudes. New, fascinating-to-put-tegether, low-cost products would serve as the vehicle to spread that love.

Again the organization would be small and engineering-oriented, staffed by eager enthusiasts, who would bust their backs to implement the basic idea.

My crystal ball gives no hint of the field involved, but I believe the commonality would include strong desires to:

Use local, or at least national, resources (raw materials, custom parts, sub-assemblies)

Keep the organization simple, with a minimum of staff functions

Use direct (mail or its successor) distribution for lowest cost to the customer

Avoid credit sales, using cash-with-order, and discounting all invoices

Deal with the customer as a friend, giving and accepting friendly help

Rely on customer goodwill and acceptance for sustained growth

I personally feel that any product with significant assembly labor content may have potential for a good kit product - especially if the market is small enough to make automated assembly unlikely.

Much has changed in this world since Ed Heath flew his first airplane and Howard Anthony built his first radio. But the joy of doing something useful and constructive, and sharing that joy with one's fellow man, are lasting concepts, I believe.

Frankly, I'm looking forward to the next implementation of the Heathkit idea. May history repeat itself!

Bjorn Heyning, 1986

A few product highlights...

1947 Oscilloscope and vacuum tube voltmeter

1952 First high fidelity kits

1955 Amateur radio products

1956 Geiger counter

1957 First transistor radio; automotive and marine products

1959 Chippewa - First KW linear; CB 'lunchbox'

1962 Black and white television

1963 Electronic organ

1965 Color television

1967 Scientific instrumentation

1968 Radio-controlled models; first home security

1969 Boonie-bike; photography products

1971 Microwave oven and trash compactor

1972 Desktop calculator

1973 Pocket calculators; Digital tuner; Weather station

1976 TV video game

1977 Digital computer products; home pinball

1978 Pocket navigational computer

1980 Projection TV; H89 'All-in-One' computer

1981 Satellite earth station

1982 Most Accurate Clock

1983 HERO 1 robot

1985 Reflector telescope

1986 HERO 2000 advanced robot

1991 Intelligent antenna rotator (last ham product)